# TRENDS IN CRYSTAL GROWTH RESEARCH

# TRENDS IN CRYSTAL GROWTH RESEARCH

**GEORGE V. KARAS**
**EDITOR**

**Nova Science Publishers, Inc.**
*New York*

For permission to use material from this book please contact us:
Telephone 631-231-7269; Fax 631-231-8175
Web Site: http://www.novapublishers.com



**LIBRARY OF CONGRESS CATALOGING-IN-PUBLICATION DATA**

Trends in crystal growth research / George V. Karas (editor).
p. cm.
Includes bibliographical references and index.
ISBN 1-59454-541-3
1. Crystal growth--Research. 2. Electronics--Materials. I. Karas, George V.
QD921.T696 2005
548'.5'072--dc22 2005014882

*Published by Nova Science Publishers, Inc. ✣ New York*

# CONTENTS

# PREFACE

Experimental and theoretical aspects of crystal growth and its applications, e.g. in devices, are within the scope of this new book. Experimental and theoretical contributions are included in the following fields: theory of nucleation and growth, molecular kinetics and transport phenomena, crystallization in viscous media such as polymers and glasses; crystal growth of metals, minerals, semiconductors, superconductors, magnetics, inorganic, organic and biological substances in bulk or as thin films; molecular beam epitaxy, chemical vapor deposition, growth of III-V and II-VI and other semiconductors; characterization of single crystals by physical and chemical methods; apparatus, instrumentation and techniques for crystal growth, and purification methods; multilayer heterostructures and their characterisation with an emphasis on crystal growth and epitaxial aspects of electronic materials.

In chapter 1, the study of pulsed-laser deposition (PLD) of $LiNbO_3$ films on sapphire and diamond substrates and the structural characterization were carried out. By controlling oxygen pressure during PLD, orientation controllable deposition of epitaxial $LiNbO_3$ films on sapphire substrates has been achieved. The structure of $LiNbO_3$ films on sapphire substrates with different orientation relationships grown under different conditions were examined by x-ray diffraction and transmission electron microscopy, and the mechanism for different growth orientations has been discussed. In addition, a two-step growth was developed to grow surface-flat c-oriented $LiNbO_3$ epitaxial films on sapphire substrates. The developed two-step growth technique was implemented for $LiNbO_3$ film growth on polycrystalline diamond substrates, and highly c-oriented $LiNbO_3$ films have been achieved by introducing a pulsed-laser deposited aluminum oxide ($Al_2O_3$) buffer layer. Gibbs free energy was employed to interpret the highly c-oriented $LiNbO_3$ films on polycrystalline diamond substrates.

Microchip laser is the most compact laser diode (LD) pumped solid-state laser with gain medium less than 1mm and can oscillate in a single longitudinal mode, so compared with the conventional solid-state laser, higher concentration of active ions is necessary for full absorption of the pumping power in such a thin microchip. However, the difficulty of crystal growth and the effect of fluorescence quenching increase more or less with the concentration of active ions doped in a crystal. In chapter 2, taking into account the crystal growth and the fluorescence quenching, some laser crystals, such as $Nd^{3+}$ and $Yb^{3+}$ doped YAG, $KGd(WO_4)_2$, and $YVO_4$, and the stoichiometric neodymium crystals are evaluated as microchip laser media. The chemical structure reasons for weak concentration dependent fluorescence quenching in the borate laser crystals are discussed. The microchip laser crystal growth, including stoichiometric laser crystal $YbAl_3(BO_3)_4$, and crystals of $YAl_3(BO_3)_4$ and $GdAl_3(BO_3)_4$ doped with series of $Yb^{3+}$ concentrations, are analyzed. Two new ways for conquering the difficulty of microchip processing, the liquid phase epitaxy and the adoption of cleavage plates, are introduced.

Relaxor ferroelectric $(1\text{-}x)Pb(Mg_{1/3}Nb_{2/3})O_3\text{-}xPbTiO_3$ (PMN-PT) single crystals have come into prominence due to their extrahigh electromechanical coupling factor ($k_{33} \sim 94\%$), piezoelectric coefficient ($d_{33} \sim 2500$ pC/N), dielectric constant ($\varepsilon_r \sim 5000 \sim 5500$) and field-induced strain response ($S \sim 1.7\%$). Many techniques have been used to grow PMN-PT single crystals, such as flux growth (high temperature solution method), melt growth (high-pressure Czochralski and Bridgman growth) and hydrothermal growth. Though most $PbTiO_3$ (PT)-based single crystals can be grown from a PbO flux, it is difficult to grow large-size and good-quality PMN-PT single crystals in a production-viable and cost-effective way. On the basis of the study of the phase stability and phase structure, the feasible method is investigated and it is verified by our researches that PMN-PT single crystals can be grown directly from melt for their stability of perovskite phase at high temperature, so a modified Bridgman technique is used to grow PMN-PT single crystals. By the B-site precursor synthesis method and using seed crystals, the crystal boules with 50 mm in diameter and about 80 mm in length have been obtained in our group. A mass of studies had been carried out on their unusually large piezoelectric response, while in chapter 3 the authors report their up-to-date optical properties. From the results of the refractive indices, the optical transmittance, the extinction ratio and the linear electro-optic (EO) properties, we have found that PMN-PT single crystals are very promising candidates for EO materials.

NiTi shape memory alloys (SMAs) are presently successful in engineering and medical technology. They combine good structural and functional properties with good corrosion resistance and biocompatibility as discussed in chapter 4. It is well known that shape memory properties of NiTi SMAs strongly depend on the processing technique and the associated microstructure. But the elementary processes which govern the chemical composition of the alloy during melting and the formation of microstructure during solidification are not yet fully understood. In the present study we investigate the formation of titanium carbide (TiC, NaCl-type B1 crystal structure) which forms during an important ingot metallurgy processing route, where NiTi SMAs are produced from pure elements using vacuum induction melting (VIM) in graphite crucibles. This procedure causes some carbon contamination of the alloy which can be tolerated up to a certain limit (e.g. industrial standard: 0.05 wt.%). We use optical microscopy (OM), X-ray diffraction (XRD), scanning electron microscopy (SEM) with energy dispersive X-ray (EDX) spectrometry, and transmission electron microscopy (TEM) to study the morphology of the carbides and to establish how microstructures form and whether or not there are specific orientation relationships between the NiTi parent phase (B2) and TiC. The solidification behaviour of carbon-containing NiTi melts is discussed on the basis of a quasi binary NiTi-TiC phase diagram. Only eutectic TiC phase was found for NiTi melts with low carbon contents. Small (regular octahedral and dendritic) and large (irregular) primary carbides appear in the microstructure of NiTi alloys as the carbon content increases. The study also shows that <100> is a preferred growth direction of eutectic and small primary TiC carbides during solidification. In addition, TEM results suggest that there exists a specific orientation relationship between the eutectic TiC phase and the B2 matrix.

Oxide single crystals are usually grown by Czochralski method. However, the vertical Bridgman method has become an important technique for industrial growth of oxide crystals in recent years. Chapter 5 reports that the vertical Bridgman method has several advantages for growing piezoelectric crystals. First, the crystal shape depends on the shape of the crucible; which means high utility of as-grown crystals during processing standard wafers. Second, the multi-crucible Bridgman furnace has been developed for mass-production of piezoelectric oxide crystals with lower cost. Third, a small temperature gradient in the solid-liquid interface and in situ annealing process considerably reduce the thermal stress in as-grown crystals. Lithium tetraborate crystal $Li_2B_4O_7$ (LBO) is a superior substrate material for surface acoustic wave (SAW) and bulk acoustic wave (BAW) devices. The Bridgman growth of 3 and 4 inches LBO crystals is presented in detail and mass production of large size LBO piezocrystals has been realized in our institute. Novel piezoelectric $Sr_3Ga_2Ge_4O_{14}$ (SGG) crystal up to 2 inches in diameter has been grown successfully by the modified vertical Bridgman method for fabrication of SAW devices. Relaxor ferroelectric crystals, such as, (1-x)$Pb(Mg_{1/3}Nb_{2/3})O_3$-x$PbTiO_3$ (PMN-PT) and (1-x)$Pb(Zn_{1/3}Nb_{2/3})O_3$-x$PbTiO_3$ (PZN-PT), possess an extremely large electric-field-induced strain (>1%) and a very high electromechanical coupling coefficient $k_{33}$ (>90%), which make them the next-generation electromechanical transducer materials in advanced applications of actuators, medical ultrasound imaging and underwater communication. The compositional inhomogeneity of crystal wafers and the erosion of high temperature solution on platinum crucible are the main problems in Bridgman growth of these lead-based ferroelectric perovskite crystals. PbO flux has been induced to Bridgman growth of PZN-PT crystals and several auxiliary techniques have been developed to improved crystal quality.

As reported in chapter 6, carbon forms various allotropes from amorphous carbon to recently-discovered fullerenes and carbon nanotubes. In spite of the elemental identity, carbon allotropes show different, albeit similar in many respects, physical properties as a result of their structural dissimilarities. Electron energy loss spectroscopy has been one of the most important tools used for measurements of their electronic properties because it can show the elementary excitations of solids in a wide energy range with a momentum transfer or with a high spatial resolution. In this study, previous reports on plasmon and core excitations of carbon allotropes investigated by electron energy loss spectroscopy are reviewed. Therefore, the positions of their $\pi$ , $\sigma$, $\pi + \sigma$ plasmons, and intraband transitions along with carbon 1s core level absorptions are compared and the reasons for similarities and differences in these excitations and for conflicts between previous reports are discussed. Furthermore, dependence of the plasmons on momentum transfer and spatial locations are reviewed as well as dielectric functions and joint energy density of states of the carbon materials. Also, for comparison electron energy loss spectra in multiwalled carbon nanotubes and amorphous carbon were recorded and assessed. Collections of spectra were recorded by a Gatan PEELS spectrometer attached to transmission electron microscopy by scanning an electron probe across carbon nanotubes or positioning it on an amorphous carbon film.

In chapter 7, aqueous solution of Urea is used as seeding agent for ice nucleation in a walk-in cold chamber. Ice nucleation is studied over a range of -19.0 °C to 0 °C. The variation of crystal concentration with temperature has a sharp peak at -11.9 °C for a specific Total Water Content (TWC). Incidentally, -11.9 °C coincides with the eutectic temperature of Urea and water mixture. When seeding is done by aqueous solution of Urea, there is a temperature drop and it has a variational pattern with seeding temperature, which is in conformity with the variation of crystal count. At fixed temperature ice crystallization depends on TWC and the number count attains a saturation value with rise in TWC. The aqueous solution of Urea has the ability to form hydrate crystal on the positive side of 0 °C up to +15.2°C. The maximum nucleation is found to occur at +11 °C. The hydrate nature of the crystal is confirmed after carrying out different supporting experiments. Hydrate crystals are grown in an incubator within the above temperature range. X-ray Diffraction (XRD) analysis establishes the presence of Urea as well as hydrated form in the grown crystal. Lastly, $^{1}H$-NMR study confirms the presence of Urea and water in the grown crystal.

In: Trends in Crystal Growth Research
Editor: George V. Karas, pp. 1-21

ISBN 1-59454-541-3


*Chapter 1*

# PULSED-LASER DEPOSITION OF $LiNbO_3$ FILMS ON SAPPHIRE AND DIAMOND SUBSTRATES

*H. K. Lam and J. Y. Dai**

Department of Applied Physics and Materials Research Centre
The Hong Kong Polytechnic University, Hung Hom, Kowloon,
Hong Kong, P. R. China

## ABSTRACT

In this work, the study of pulsed-laser deposition (PLD) of $LiNbO_3$ films on sapphire and diamond substrates and the structural characterization were carried out. By controlling oxygen pressure during PLD, orientation controllable deposition of epitaxial $LiNbO_3$ films on sapphire substrates has been achieved. The structure of $LiNbO_3$ films on sapphire substrates with different orientation relationships grown under different conditions were examined by x-ray diffraction and transmission electron microscopy, and the mechanism for different growth orientations has been discussed. In addition, a two-step growth was developed to grow surface-flat c-oriented $LiNbO_3$ epitaxial films on sapphire substrates. The developed two-step growth technique was implemented for $LiNbO_3$ film growth on polycrystalline diamond substrates, and highly c-oriented $LiNbO_3$ films have been achieved by introducing a pulsed-laser deposited aluminum oxide ($Al_2O_3$) buffer layer. Gibbs free energy was employed to interpret the highly c-oriented $LiNbO_3$ films on polycrystalline diamond substrates.

## INTRODUCTION

Lithium niobate ($LiNbO_3$) is one of the most important materials for both fundamental and applied research in optics and materials science, owing to its electrical, electro-optical, nonlinear optical and crystallographic properties. As a ferroelectric crystal, $LiNbO_3$ shows a spontaneous polarization at temperature below the Curie temperature $T_c$ (1210 $^oC$). In the

* Corresponding author: apdaijy@inet.polyu.edu.hk

microscopic viewpoint, $LiNbO_3$ consists of a grid of oxygen ion ($O^{2-}$) layers in which the lithium ions ($Li^+$) and niobium ions ($Nb^{5+}$) are positioned. The displacement of the $Li^+$ and $Nb^{5+}$ relative to the neutral configuration (paraelectric phase) gives rise to a dipole moment and thus to the spontaneous polarization [1]. This is called ferroelectricity, in which the c-axis of $LiNbO_3$ is the direction of this displacement and thus is the ferroelectric axis of $LiNbO_3$.

Started in the 1970s, surface acoustic wave (SAW) devices have become attractive as components incorporated in electronic systems [2]. One possible approach to improve the device performance is to widen the choice of substrate materials by using piezoelectric films with interdigital transducers (IDTs) to generate and detect surface waves on nonpiezoelectric substrates. $LiNbO_3$ has been deposited on different substrates or combined with another piezoelectric layer to enhance the performance of SAW devices. For the $LiNbO_3$ film deposition, it is usually done on the sapphire substrate [3-5], since these materials are in hexagonal structure and their lattice mismatch are not so large (along the c-axis is 6.7%, while along the a-axis is 8.0%). By controlling the oxygen pressure during pulsed laser deposition (PLD), orientation controllable deposition of $LiNbO_3$ films by PLD can be achieved.

For the successful deposition of epitaxial $LiNbO_3$ films on sapphire, a number of different growth methods have been reported, primarily by magnetron sputtering, molecular beam epitaxy and most recently by the sol-gel process and PLD [6]. Although PLD gives rise to stoichiometric films, the formation of single or textured $LiNbO_3$ films is still difficult to be achieved since Li is highly volatile. Shibata and co-workers succeeded in using Li enriched target (Li/Nb ratio = 2) to deposit epitaxial $LiNbO_3$ films on the (0001) sapphire substrates at high temperature and low oxygen pressure ambient by PLD [3-4]. Using a relatively low temperature approach, Lee and co-workers successfully deposited stoichiometric $LiNbO_3$ films by PLD on sapphire substrates using single crystal $LiNbO_3$ target [7]. Both a- and c-oriented $LiNbO_3$ films have been deposited on a- and c-sapphire substrates, and the mechanism of the orientation controllable growth has been studied. However, only oxygen pressure was considered as the controlling factor on the $LiNbO_3$ film orientation in their work.

Besides the application in SAW devices, there are several other reasons suggested for the investigation of $LiNbO_3$ films. One is the application in optical devices. Heteroepitaxy of $LiNbO_3$ on substrates such as sapphire or silicon would appear to offer advantages for optical devices [8-12], for example, improved electro-optic (EO) properties may be achieved [13]. Since the EO properties of $LiNbO_3$ are anisotropic, the desired epitaxial growth of films can reduce the light scattering due to a refractive index mismatch at grain boundaries [14]. The EO effect can be modulated by different epitaxial growth orientations and dopants of $LiNbO_3$. The possibilities of producing periodically poled $LiNbO_3$ and the fabrication of multilayer structures also are the great advantage of $LiNbO_3$ films [15]. The advanced applications of $LiNbO_3$ make it an attractive candidate in many research areas.

$LiNbO_3$ exhibits various kinds of prominent properties, such as piezoelectric, electro-optic, photo-elastic and nonlinear optical effects, and for different applications, different crystal orientations of $LiNbO_3$ films may be needed. For example, <0001> oriented $LiNbO_3$ films are required in synthesizing SAW devices in order to fully utilize the $d_{33}$ piezoelectric effect. By contrast, electro-optical modulator devices require the $LiNbO_3$ films to be <0001> in-plane oriented so that the in-plane piezoelectric property can be used to modulate the refractive index. Therefore, it is worthy to study the detailed structural properties of $LiNbO_3$ films.

Recently, $LiNbO_3$/diamond is becoming an attractive combination in SAW device fabrication to meet the demand of high frequency devices, since diamond substrate is a possible candidate to achieve the GHz range SAW device. Magnetron sputtering deposition of $LiNbO_3$ films on diamond substrates has been reported earlier [16-18]. The $LiNbO_3$ films are randomly oriented on diamond/Si substrates, and due to the arbitrary crystal orientation of $LiNbO_3$ films on diamond, the insertion loss of the SAW devices utilizing the $LiNbO_3$/diamond structure is relatively high [19-20]. To the best of our knowledge, there is no report on successful deposition of highly oriented $LiNbO_3$ film on diamond substrate. The arbitrary crystal orientation of $LiNbO_3$ films on diamond substrates may be due to the polycrystalline structure of the diamond substrate. Another possible reason is that the polycrystalline diamond is easy to be oxidized at high temperature in oxygen ambient. This makes the deposition of highly c-oriented and surface-smoothed $LiNbO_3$ films difficult [16]. It is believed that a buffer layer is needed to protect the diamond substrates from oxidation. Aluminum oxide ($Al_2O_3$) is a promising buffer layer due to its relatively high acoustic velocity, high thermal conductivity and good oxygen diffusion barrier properties [21]. During the $LiNbO_3$ film deposition, the $Al_2O_3$ layer may prevent the diamond substrate from oxidation at high temperature and high oxygen pressure environment. Furthermore, Hu and co-workers reported that the $Al_2O_3$ layer can enhance the <0001> direction growth of $LiNbO_3$, which makes it possible to grow highly c-oriented films rather than one with random orientation [22].

In this work, we mainly focus on the growth of $LiNbO_3$ films by PLD on sapphire and diamond substrates and their structural characterization. The optimized deposition conditions and growth mechanism of $LiNbO_3$ films on the sapphire and diamond substrates were investigated. We also proposed a more sophisticated model to explain the $LiNbO_3$ grwoth in considering adatom kinetic energy and Gibbs free energy.

# PART 1. $LiNbO_3$ ON SAPPHIRE

## 1.1. Growth of $LiNbO_3$ Films

Conditions for the epitaxial growth of $LiNbO_3$ films on the (0001) (c-oriented) and $(11\bar{2}0)$ (a-oriented) sapphire substrates have been investigated by varying the PLD growth conditions such as oxygen pressure and substrate temperature. The $LiNbO_3$ films were deposited by PLD with KrF excimer laser ($\lambda = 248$ nm) at 250 mJ and 5 Hz repetition rate. A $LiNbO_3$ single crystal target was used in the deposition. The deposition temperature for the $LiNbO_3$ films was at 600-650 °C. Three approaches with different oxygen pressures have been used in $LiNbO_3$ film deposition on c-oriented substrates, i.e., 500 mTorr, 100 mTorr, and two-step growth (a seed layer deposition at 500 mTorr for 15 s followed by 12 min growth at 100 mTorr). While on a-sapphires, 500 mTorr, 100 mTorr and 10 mTorr oxygen pressures have been used. After the deposition, the films were in-situ annealed at 650 °C for 30 min in oxygen pressure of 11 Torr. The same pressure was maintained until the sample has been cooled down to room temperature.

The surface morphology of $LiNbO_3$/sapphire was observed by field-emission scanning electron microscope (SEM). X-ray diffractometry (XRD) and transmission electrons microscopy (TEM) were used to investigate the structural properties, and the interfacial structure of the films at different oxygen pressures. The electronic structure and dielectric function of $LiNbO_3$ films were obtained from the electron-energy loss spectrum (EELS).

## 1.2. Structural Characterization and Surface Morphology

### *1.2.1. $LiNbO_3$ Films on C-Sapphire Substrates*

Crystal orientation analysis by XRD revealed that 500 mTorr pressure resulted in c-oriented $LiNbO_3$ film growth in both c- and a-sapphire substrates even though the c-$LiNbO_3$ and a-sapphire possess different crystal symmetry. This result is similar to Lee and co-worker's reports [7, 14] where a relatively fair explanation has been given. However, 100 mTorr of oxygen pressure are not favorable in growing c-oriented $LiNbO_3$ films even on c-orientated sapphire substrates, therefore, the epitaxial growth of surface-flat $LiNbO_3$ on sapphire substrates with lower oxygen pressure is difficult. It is reasonable to expect, that if a seed layer of c-oriented epitaxial $LiNbO_3$ can be formed with 500 mTorr growth, a second step growth at lower oxygen pressure of 100 mTorr may keep the c-oriented $LiNbO_3$ growth with improved surface roughness. Based on this, we proposed a two-step growth method, i.e., a seed layer deposition at 500 mTorr followed by a growth at 100 mTorr.

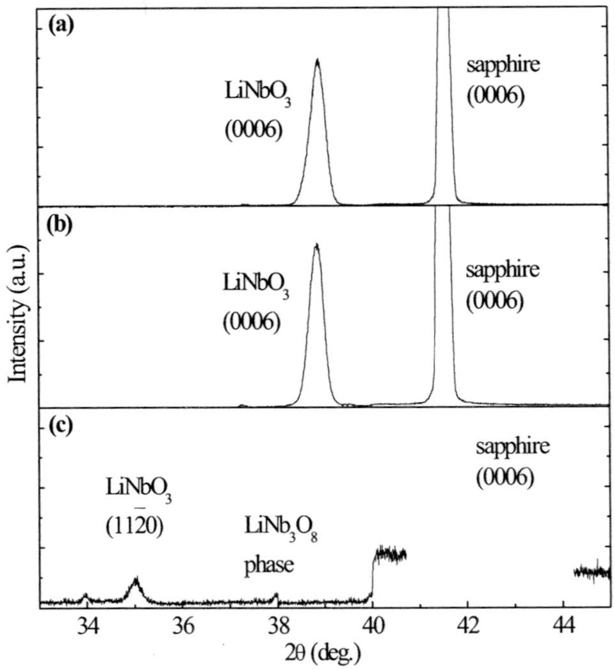


Figure. 1.1 XRD patterns of the $LiNbO_3$ films grown on c-sapphire substrates at 650 °C in different oxygen pressures: (a) 500 mTorr, (b) two-step method (500 mTorr for 15 s followed by 10 min at 100 mTorr), and (c) 100 mTorr.

Figure 1.1 shows XRD patterns of $LiNbO_3$/c-sapphire obtained under various oxygen pressures, i.e., 500 mTorr, two-step (500 mTorr for 15s followed by 10 min at 100 mTorr ) and 100 mTorr. All $LiNbO_3$ films in this figure were grown at 650 °C. As shown in Figs. 1.1 (a) and (b), the films deposited with 500 mTorr and the two-step method show a strong (0006) reflection of $LiNbO_3$. By contrast, under low oxygen pressure deposition (100 mTorr), the $(11\bar{2}0)$ reflection of $LiNbO_3$ and Li deficient phase ($LiNb_3O_8$) appear (see Fig. 1.1 (c)). It revealed that both high oxygen pressure (500 mTorr) and the two-step growth conditions favor the growth of c-oriented $LiNbO_3$ films on c-sapphire substrates.

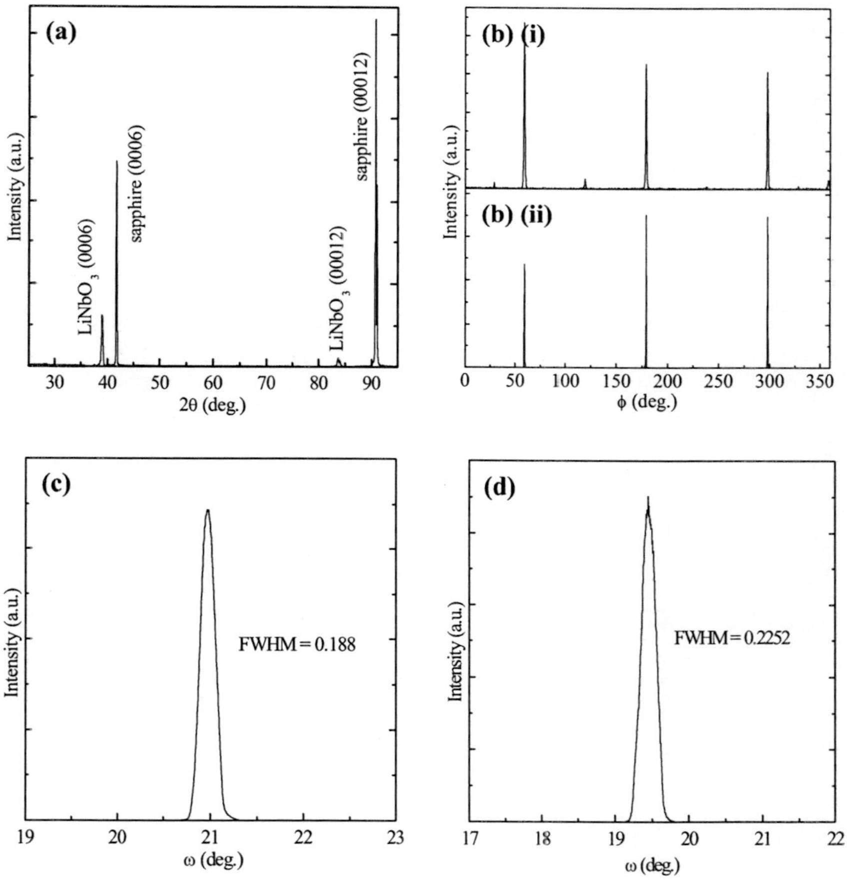


Figure. 1.2. (a) XRD patterns of LiNbO3/c-sapphire grown at 650 oC by the two-step growth, (b) the $\phi$ scans of LiNbO3/c-sapphire: (i) (0006) reflection of the LiNbO3 film, and (ii) (0006) reflection of sapphire substrate, (c) the $\omega$ scan of sapphire (0006) reflection, and (d) the $\omega$ scan of (0006) reflection of the LiNbO3 film.

More detailed study by XRD of the $LiNbO_3$ films on c-sapphire revealed the epitaxial growth for both of the films grown under 500 mTorr and the two-step method. Figure 1.2

shows the detailed structural investigation of $LiNbO_3$/c-sapphire obtained by the two-step growth method. As shown in Fig. 1.2 (a), the $\theta - 2\theta$ scan of the $LiNbO_3$/c-sapphire under the two-step growth condition shows strong reflections of $LiNbO_3$ (0006) and (00012) only. It illustrates that the $LiNbO_3$ film is highly c-oriented without the presence of the Li deficient phase inside the film. Figure 1.2 (b) also indicates that the film is epitaxially grown on the c-sapphire substrate. The three-fold symmetry of the film is exactly the same as that of the c-sapphire substrate. In Figs. 1.2 (c) and (d), the rocking curve of the (0006) reflection for the $LiNbO_3$ thin film shows the full width half maximum (FWHM) of $0.22^{\circ}$. This value is close to the FWHM relative to the sapphire (0006) reflection ($0.19^{\circ}$). This confirms that the c-oriented $LiNbO_3$ film has high degree of crystallinity on the c-sapphire substrate.

However, the surface flatness was found to be greatly affected by the oxygen pressures. Figure 1.3 (a) shows the SEM image of $LiNbO_3$/c-sapphire deposited at 650 °C in 500 mTorr oxygen pressure. It reveals that there are a lot of particulates on the surface of the film. The film with such a high surface roughness can not be applied for the fabrication of SAW devices even though the crystallization of the film is good at this growing condition. For the films grown by the two-step method and 100 mTorr (see Figs. 1.3 (b) and (c)), the particulate problem was eliminated, resulting in a low surface roughness of films. The surface roughness grown by the two-step method is comparable with that grown at 100 mTorr.

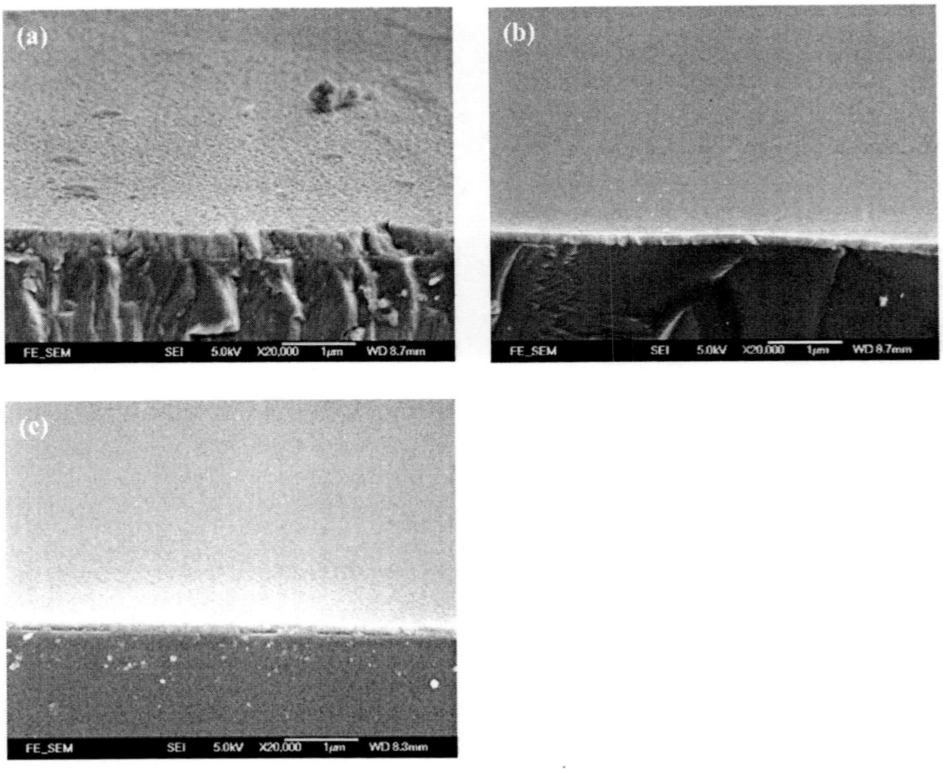


Figure 1.3. SEM images of $LiNbO_3$/c-sapphire grown at 650 °C under different oxygen pressures: (a) 500 mTorr, (b) two-step method (500 mTorr for 15 s followed by 10 min at 100 mTorr), and (c) 100 mTorr.

During the deposition of $LiNbO_3$ films, cracks would be formed as the film thickness increase. In most of the films with thickness of more than 600-700 nm, cracks and peeling off of the film can be observed clearly on the film surfaces (see Fig. 1.4). These cracks may be caused by the mismatches of the thermal expansion coefficients or lattice constants between the $LiNbO_3$ films and sapphire substrates. A similar result has also been reported earlier by Shibata and co-workers [3].

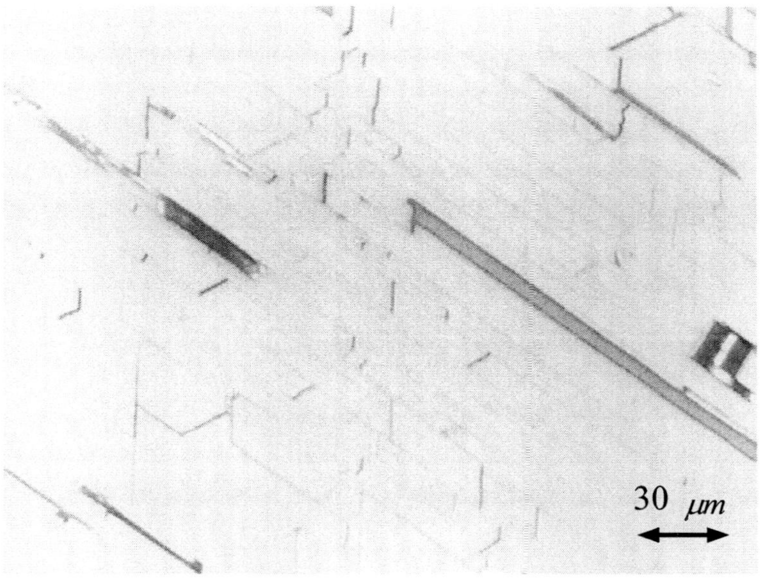


Figure 1.4. Optical micrograph of 600 nm-thick $LiNbO_3$ film on c-sapphire substrate (magnification of $10\times 10$).

### *1.2.2. $LiNbO_3$ Films on a Sapphire Substrates*

Figure 1.5 shows the XRD patterns of $LiNbO_3$/a-sapphire deposited at 650 °C under different oxygen pressures, i.e., 500 mTorr, 100 mTorr and 10 mTorr. The $\theta - 2\theta$ scan, as shown in Fig. 1.5 (a), illustrates that c-oriented $LiNbO_3$ films were grown on a-sapphire substrates under 500 mTorr deposition oxygen pressure. It shows that high oxygen pressure deposition also favors c-oriented growth of $LiNbO_3$ films on a-sapphire substrates, even their lattice mismatch is much larger. By contrast, at low oxygen pressure, the growths of a-oriented $LiNbO_3$ films become favorable. Both of Figs. 1.5 (b) and (c) show a $(11\bar{2}0)$ reflection of $LiNbO_3$ in $\theta - 2\theta$ scan, without any (0006) reflection peak of $LiNbO_3$. However, at 10 mTorr oxygen pressure, Li deficient phase ($LiNb_3O_8$) presents inside the film. It can be concluded that epitaxial a-oriented $LiNbO_3$ films on a-sapphire can be obtained under 100 mTorr deposition oxygen pressure, while at an oxygen pressure much lower than this value, Li deficient phase will be formed.

Figures 1.6 and 1.7 show the more detailed structural investigation of the $LiNbO_3$/a-sapphire under 500 mTorr and 100 mTorr depositions. Figure 1.6 (a) shows a very sharp (0006) $LiNbO_3$ and $(11\bar{2}0)$ sapphire reflections. In the $\phi$ - scan, the $LiNbO_3$ film exhibits a 3-fold symmetry with two structural variants (twined crystals) which are rotated by 60° about their c-axis, while the a-sapphire substrate exhibits a 2-fold symmetry (see Fig. 1.6 (b)). This twinning has often been reported in the literature and was attributed to the lattice asymmetry of two-to-three fold between $LiNbO_3$ and sapphire [8, 10, 23]. The FWHM of the rocking curve of the $LiNbO_3$ (0006) is 0.51°, while it is 0.13° for sapphire $(11\bar{2}0)$ peak (see Figs. 1.6 (c) and (d)).

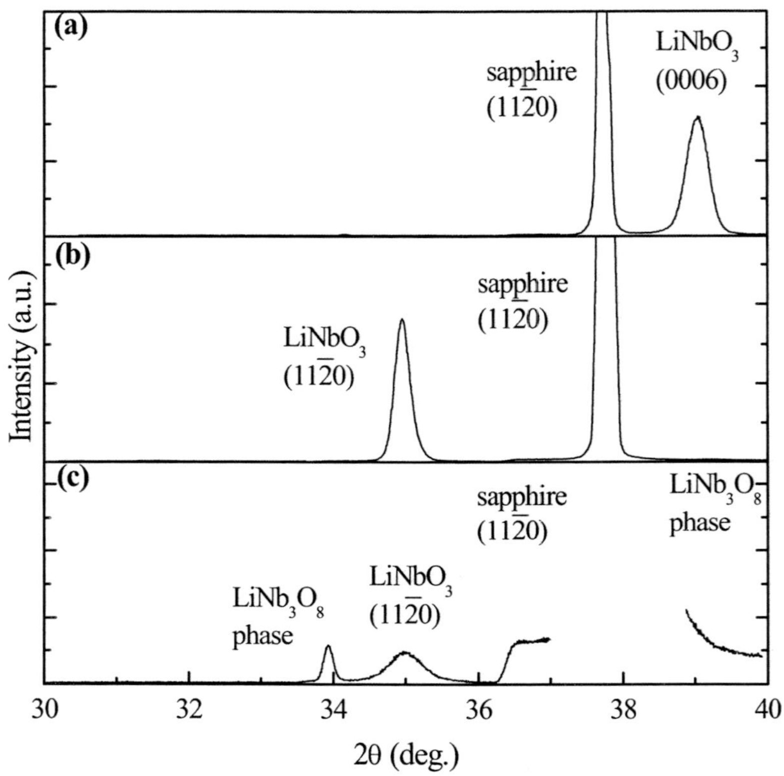


Figure 1.5. XRD patterns of the $LiNbO_3$ films grown on a-sapphire substrates at 650 °C under different oxygen pressures: (a) 500 mTorr, (b) 100 mTorr, and (c) 10 mTorr.

The XRD pattern, as shown in Fig. 1.7 (a), illustrates that the $LiNbO_3$ film is highly a-oriented on the a-sapphire substrate. However, the $LiNbO_3$ film shows two more twined peaks than the 2-fold symmetry of sapphire substrate in the $\phi$ - scan. The untwined peaks appear at 168.3° and 348.1° while the twined peaks appear at 92.6° and 272.2° (see Fig. 1.7 (b)). This suggests that the a-axis of the $LiNbO_3$ film is parallel to that of the sapphire substrate and the epitaxial film is composed of twinning which are rotated by 76° about the c-axis of the substrate. In Fig. 1.7 (d), the FWHM of the rocking curve of the $LiNbO_3$ $(11\bar{2}0)$ reflection shows a value of 0.26°, which is close to that of the sapphire substrate (see Fig. 1.7 (c)). This suggests that the films are highly a-oriented on the a-sapphire substrate.

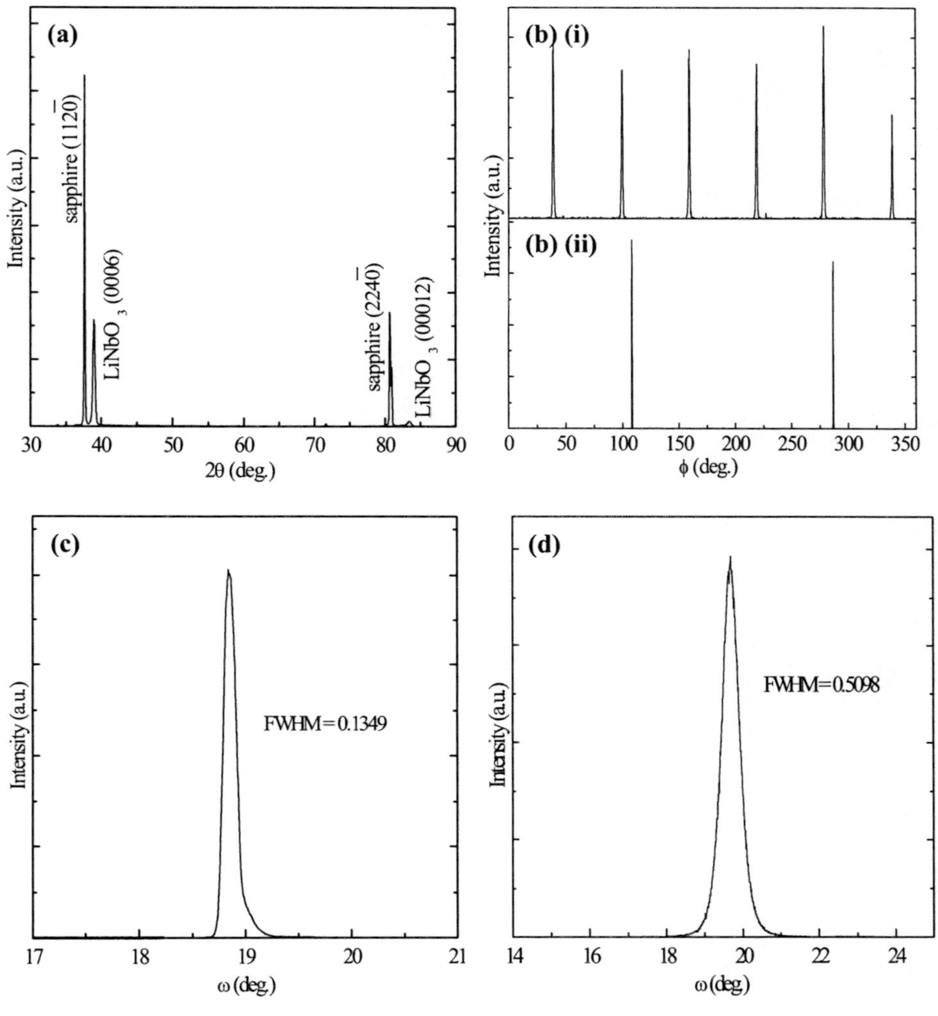


Figure 1.6. (a) XRD patterns of $LiNbO_3$/a-sapphire grown at 650 °C with 500 mTorr, (b) the $\phi$ scans of $LiNbO_3$/a-sapphire: (i) (0006) reflection of the $LiNbO_3$ film, and (ii) $(11\bar{2}0)$ reflection of sapphire substrate, (c) the $\omega$ scan of sapphire $(11\bar{2}0)$ reflection, and (d) the $\omega$ scan of (0006) reflection of the $LiNbO_3$ film.

## 1.3. Transmission Electron Microscopy (TEM) Study of the Interfacial Structure

In order to know the detailed structure of the film/substrate interface and thus to understand more of the growth mechanism, $LiNbO_3$ films on a-oriented sapphire substrate samples were carried out for a cross-sectional TEM study. The TEM sample preparation procedure includes mounting the samples by epoxy with the film side face-to-face followed by mechanical polishing and finally ion milling. Figure 1.8 (a) shows cross-sectional TEM image of the c-oriented $LiNbO_3$ on a-oriented sapphire substrate deposited at 500 mTorr of oxygen pressure, and the diffraction pattern, as shown in Fig. 1.8 (b), illustrating the orientation relationship between the $LiNbO_3$ film and the sapphire substrate, i.e. $(0001)_{LiNbO_3} // (11\bar{2}0)_{sapphire}$ and $[01\bar{1}0]_{LiNbO_3} // [01\bar{1}0]_{sapphire}$. The film exhibits a columnar and epitaxial growth feature with grain size of about 50 nm. Figure 1.8 (c) is the corresponding HREM image showing the interfacial structure. It can be seen that the interface is free from interfacial reaction, and the interface is sharp in spite of the presence of a large number of misfit dislocations at the interface due the lattice mismatch.

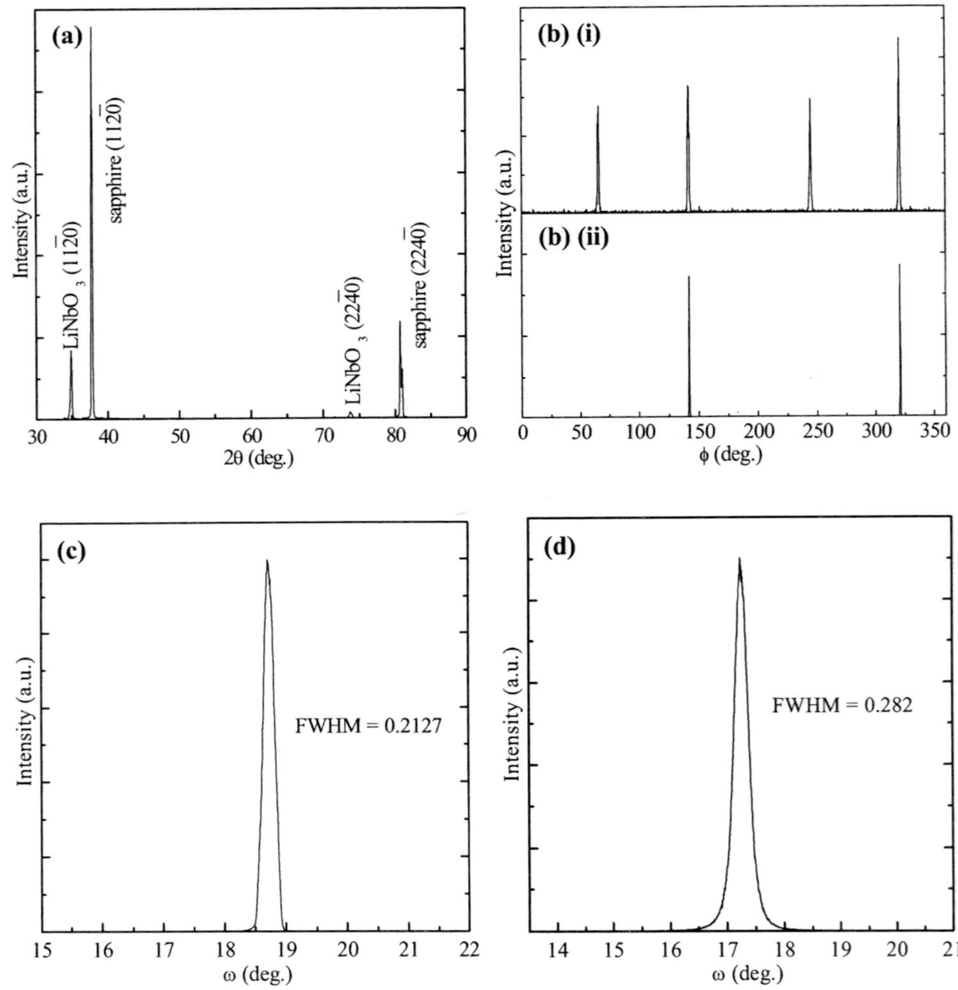


Figure 1.7. (a) XRD patterns of $LiNbO_3$/a-sapphire grown at 650 °C with 100 mTorr, (b) the $\phi$ scans of $LiNbO_3$/a-sapphire: (i) $(11\bar{2}0)$ reflection of the $LiNbO_3$ film, and (ii) $(11\bar{2}0)$ reflection of sapphire substrate, (c) the $\omega$ scan of sapphire $(11\bar{2}0)$ reflection, and (d) the $\omega$ scan of $(11\bar{2}0)$ reflection of the $LiNbO_3$ film.

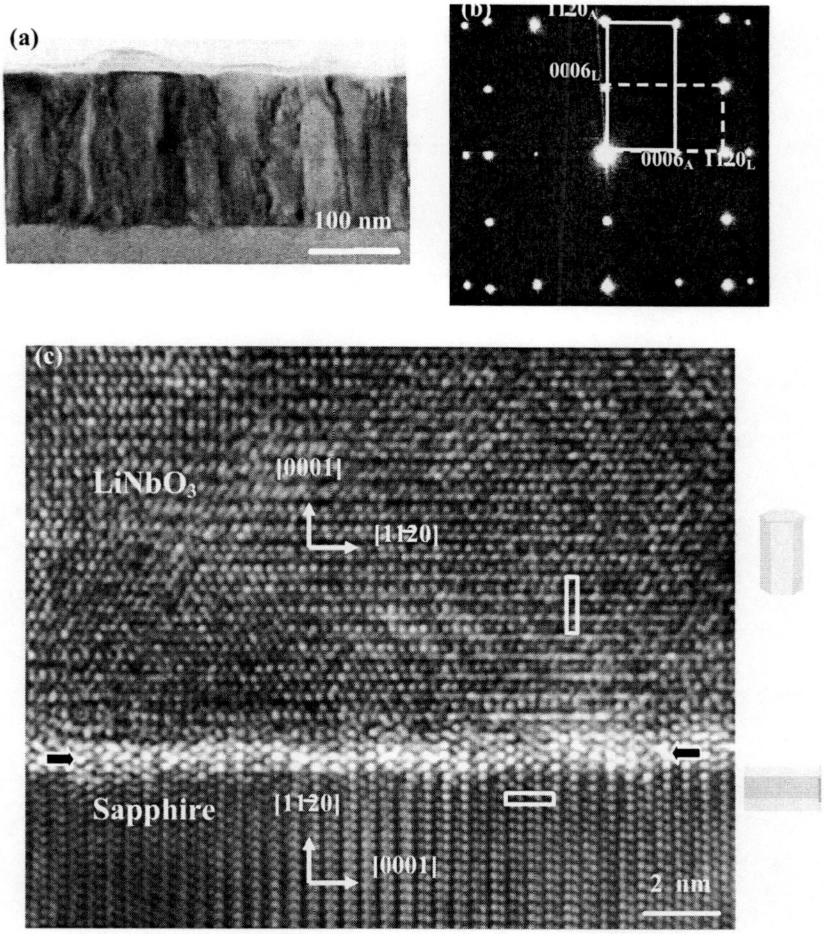


Figure 1.8. (a) Cross-sectional TEM image of c-$LiNbO_3$ on a-sapphire substrate deposited at 500 mTorr oxygen pressure, (b) the diffraction pattern showing the orientation relationship between $LiNbO_3$ film and sapphire substrate, and (c) the corresponding HREM image of the interface. Insets are the schematic diagrams of crystal structure alignment of $LiNbO_3$ film and sapphire substrate.

Figure 1.9 (a) is a TEM image showing the a-$LiNbO_3$ on a-sapphire substrate grown at 100 mTorr. The corresponding orientation relationship between the film and substrate is illustrated by the composite electron diffraction pattern as shown in Fig. 1.9 (b), i.e., $(11\bar{2}0)_{LiNbO_3} //(11\bar{2}0)_{sapphire}$ and $[01\bar{1}0]_{LiNbO_3} //[01\bar{1}0]_{sapphire}$. Again, from the corresponding HREM image as shown in Fig. 1.9 (c), it can be seen that the interface is sharp and free from interfacial reaction. The misfit dislocations exhibit periodic feature with a mean distance of about 6 nm and the induced strains at the interface can be clearly seen. The mean distance of interfacial misfit dislocations can be calculated by

$$d = \frac{d_1 \cdot d_2}{d_1 - d_2}, \qquad (1)$$

where $d_1$ and $d_2$ are distances of the mismatched atomic planes of the film and substrate, respectively. The misfit planes at the interface are 1/3(0001) plane of $LiNbO_3$ and sapphire, where $d_1 = 4.62$ Å, $d_2 = 4.33$ Å. From equation (1), the mean distance between the interfacial dislocations is 6.8 nm. However, due to the presence of a grain boundary, as indicated by an arrow, the periodicity of the misfit dislocations is interrupted.

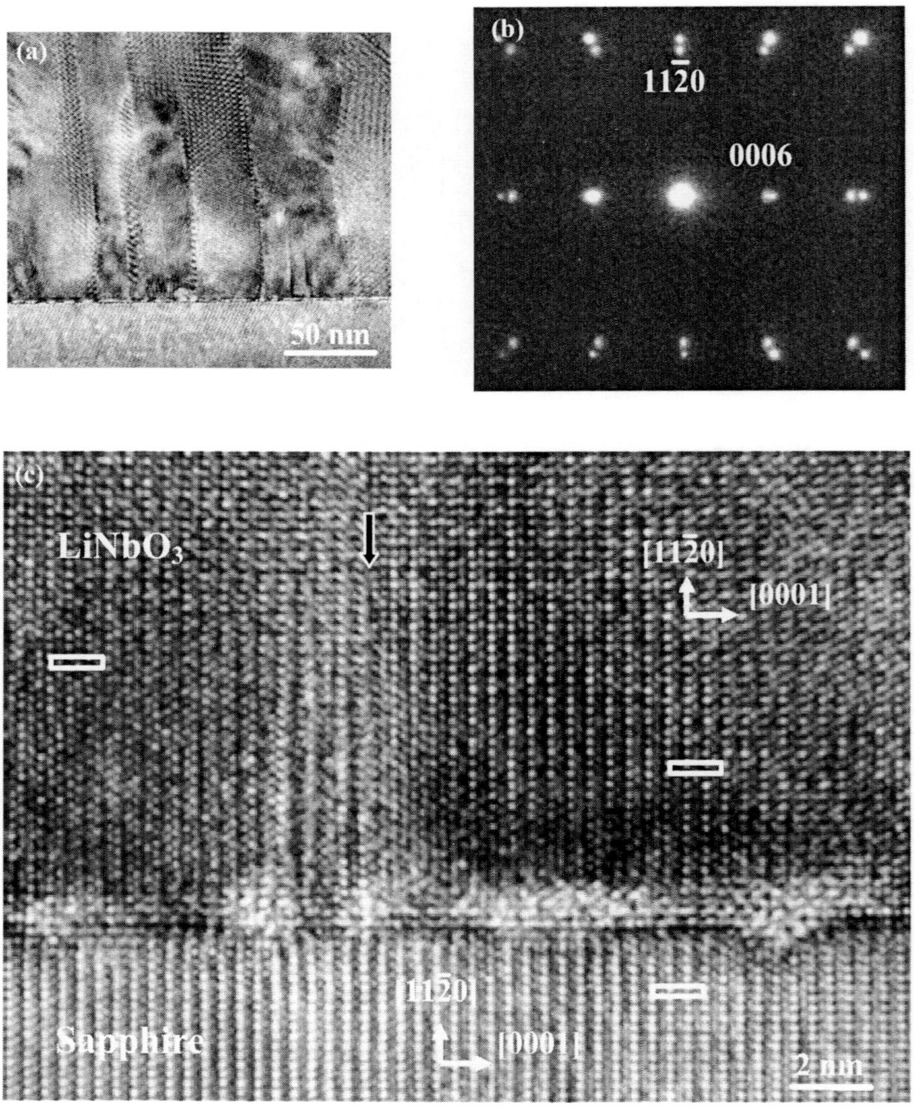


Figure 1.9. (a) Cross-sectional TEM image of c-$LiNbO_3$ on a-sapphire substrate deposited at 100 mTorr oxygen pressure, (b) the diffraction pattern showing the orientation relationship between $LiNbO_3$ film and sapphire substrate, and (c) the corresponding HREM image of the interface.

It is worthy to note that some areas in the $LiNbO_3$ films deposited at 100 mTorr oxygen pressure show co-existence of a- and c-orientations as shown in Fig. 1.10 (a). The percentage of the c-oriented grains is about 10 %. This suggests that the interfacial mismatch energies between a- and c-oriented $LiNbO_3$ with a-oriented sapphire is small and may be tolerated by other factors such as oxygen pressure, and laser beam energy etc. The film orientation does not change as the film thickness increases, but we noticed that some small c-oriented grains were finally "eaten" up by the neighboring a-oriented domains (see Fig. 1.10 (b)). This suggests that a-oriented grains may grow faster than the c-oriented grains at low oxygen pressure.

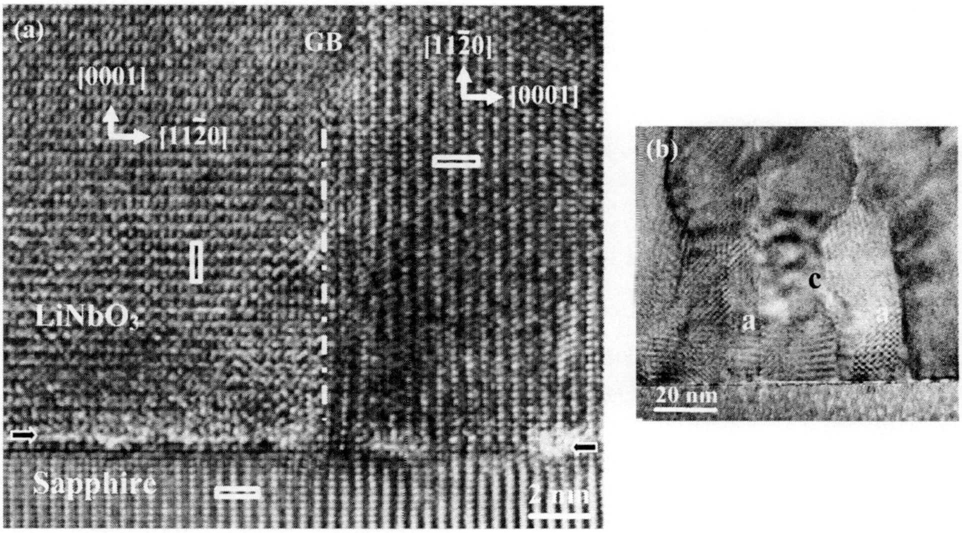


Figure 1.10. (a) HREM of the $LiNbO_3$ film on a-sapphire substrate deposited at 100 mTorr oxygen pressure showing co-existence of a- and c-oriented $LiNbO_3$. The white dot line indicates the grain boundary between a- and c-oriented $LiNbO_3$. (b) Low magnification image showing different orientation of grains.

## 1.4. Electron Energy Loss Function and Electronic Structure

EELS of the a- and c-oriented $LiNbO_3$ films were obtained using the diffraction mode at zero momentum transfer. The energy-loss functions of these two different oriented $LiNbO_3$ films after the multiple scattering corrections were obtained from the raw EELS spectra, and it turns out that the general shapes of the spectra are the same and there is no significant difference to be noticed. Figure 1.11 represents a typical energy loss function obtained from the $LiNbO_3$ films. Based on the first-principle calculation by Kohiki *et al* [24] and experimental results reported by Kasper and Hüfner [25], the electron energy losses of about 7.0 eV, 14.0 eV and the small shoulder at 11.0 eV can be attributed to electron transitions from maximum density of states in valence band to the split conduction band, where the valence band is predominantly of O 2*p* character and the conduction band contains Nb *d* and *s-p* states.

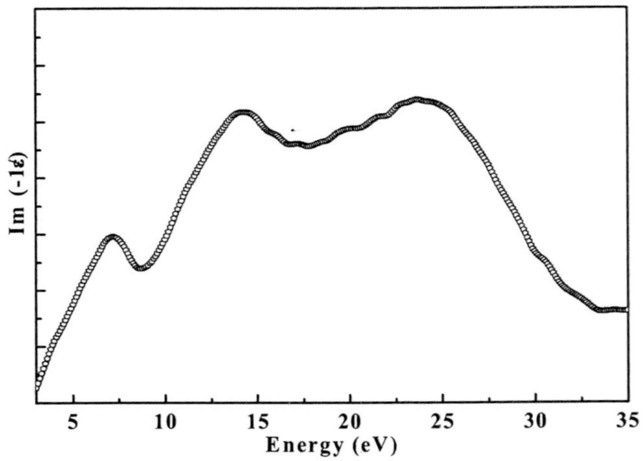


Figure 1.11. A represented energy-loss function of the $LiNbO_3$ films after the multiple scattering corrections from the raw energy–loss spectra.

Kasper and Hüfner have reported that the Li or O deficiency induced reduction of $Nb^{5+}$ to $Nb^{4+}$ and resulted in an energy loss peak at 5.2 eV. The fact that such peak was absent in our PLD deposited $LiNbO_3$ films suggests that the $LiNbO_3$ films are well crystallized without significant oxygen and Li deficiencies. The peak at 24 eV in the energy loss function is attributed to a plasmon peak.

Figures 1.12 (a) and (b) show the real and imaginary parts of the dielectric function derived from the loss function by using the Kramers-Kronig transformation. The peaks in Im(ε) correspond to the joint density of states in the valence and conduction bands, and the broad peaks at 4.0-5.0 eV and 9.0-10.0 eV match fairly well with the first-principle calculation result as reported by Kohiki *et al* [24].

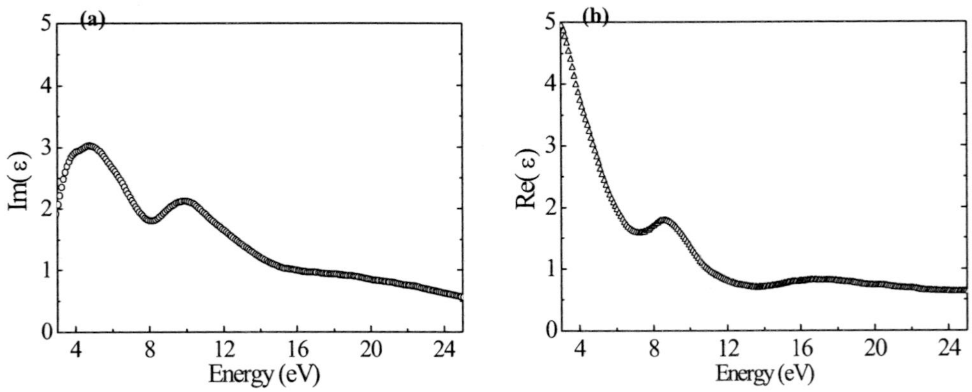


Figure 1.12. Dielectric function of the $LiNbO_3$ films derived by using Kramers-Kronig transformation: (a) Im (ε), and (b) Re(ε).

## 1.8. Growth Mechanism of $LiNbO_3$ Films

The XRD and TEM results revealed that oxygen pressure is the key parameter that affects the growth orientation of $LiNbO_3$ on sapphire substrates. Higher oxygen pressure favors the c-axis $LiNbO_3$ growth and lower oxygen pressure favors the a-axis growth. Lee and co-workers have ascribed this to the different atomic arrangements of oxygen layers in the c- and a-oriented $LiNbO_3$, i.e., at high oxygen pressure, the closely packed oxygen surface (0001) of $LiNbO_3$ is easier to grow initially on (0001) sapphire substrate [7, 14].

However, the results showed that this rule was broken if the laser energy changed. For example, if the laser fluence reduced from 5 $J/cm^2$ to 3 $J/cm^2$, the same low oxygen pressure (100 mTorr) still resulted in c-$LiNbO_3$ growth. It suggested that the adatom energy is an important factor that controls the crystal orientation of $LiNbO_3$ films during the nucleation stage, and the oxygen pressure also affects the growth orientation through changing the adatom energy. In the high oxygen pressure environment, the laser ablated materials lose their kinetic energy through the collision with the oxygen molecules. In fact, both higher oxygen pressure and lower laser fluence result in lower adatom energy, therefore, the adatom energy should play the most important role in controlling the film growth orientation.

The lattice similarity between the $LiNbO_3$ and sapphire determines that the interfacial energy of a-$LiNbO_3$/a-sapphire is lower than that of c-$LiNbO_3$/a-sapphire. However, another competition factor is the nucleation surface free energy. It is believed that when the oxygen pressure is higher, its rapid quenching effect results in lower adatom energy and the lower adatom energy makes them difficult to overcome the energy barrier to allocate to the correct lattice sites and thus the surface free energy becomes a dominant factor in controlling the growth orientation of $LiNbO_3$ nucleus. Since $LiNbO_3$ is in hexagonal structure, the (0001) lattice plane which has the largest atomic density and lowest surface free energy thus grows preferentially along the c-axis on amorphous substrates. Therefore, if the difference in the lattice mismatches between the c- and a-oriented $LiNbO_3$ with a-oriented sapphire is small, in fact this is true [7], c-oriented $LiNbO_3$ will be a preferred orientation when the adatom energy is relatively low. Based on the adatom energy understanding, we believe that the growth orientation of $LiNbO_3$ films on c-sapphire substrate may also be controlled by altering the laser fluence.

# PART 2. $LiNbO_3$ ON DIAMOND

## 2.1. Aluminum Oxide ($Al_2O_3$) Buffer Layer

$Al_2O_3$ and $LiNbO_3$ films were subsequently deposited on diamond (500 μm) substrates (supplied by Beijing Tiandi Orient Super-abrasives Materials Co., Ltd.) by pulse-laser deposition (PLD) with KrF excimer laser ($\lambda$ = 248 nm) at 250 mJ and 5 Hz repetition rate. $Al_2O_3$ thin films with thickness of 40–60 nm were grown on the polycrystalline diamond substrates with both sides polished under a base vacuum of 1.5 x $10^{-3}$ mTorr at 300 °C using an $Al_2O_3$ target. Three approaches with different oxygen pressures have been used in $LiNbO_3$ film deposition at 600 °C, i.e., 500 mTorr, 100 mTorr, and a two–step growth (growth at 500 mTorr for 15 s followed by 12 min growth at 100 mTorr), respectively. After deposition, the films were *in-situ* annealed at 600 °C for 30 min under different oxygen pressures of 11 Torr and 1 atm. The same pressure was maintained until the sample has been cooled down to room temperature.

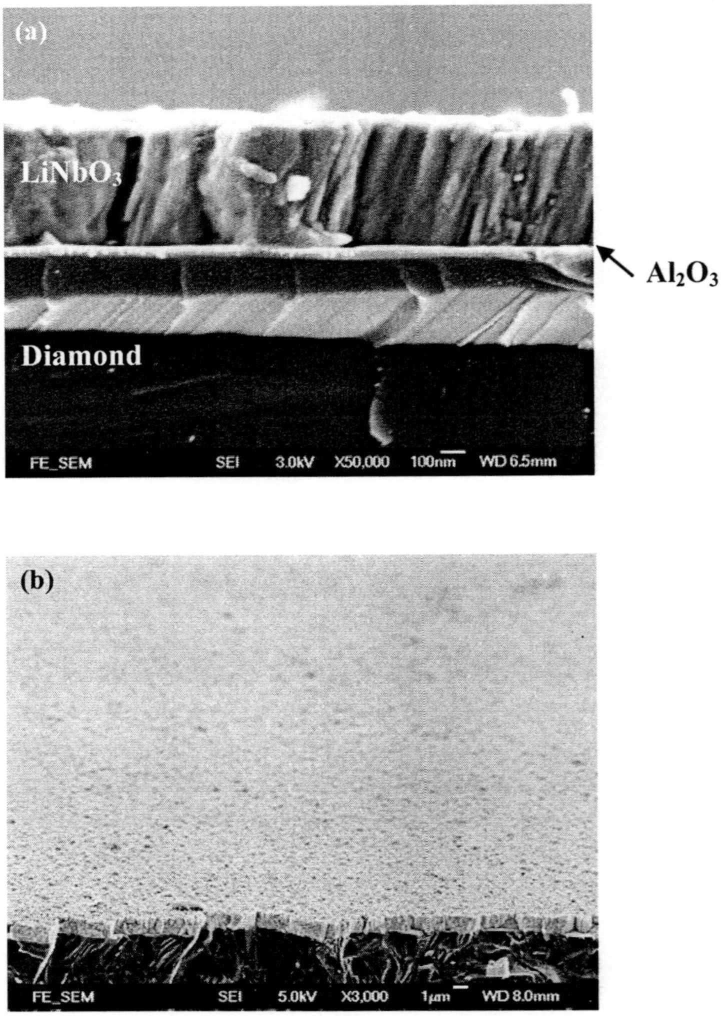


Figure 2.1. SEM images of $LiNbO_3$ film deposited on diamond substrate with $Al_2O_3$ buffer layer at 600 °C and 500 mTorr oxygen pressure: (a) cross-section, and (b) surface morphology.

## 2.2. Surface Morphology and Structural Analysis

Figures 2.1 (a) and (b) show a cross-sectional and a plane-view SEM images of the $LiNbO_3$ film on $Al_2O_3$ buffered diamond substrate. An amorphous $Al_2O_3$ buffer layer with 40 nm thickness can be clearly seen on the diamond surface. The SEM image also reveals a dense 500 nm-thick $LiNbO_3$ layer without obvious defects in both cross-section and surface morphology. In contrast, a direct deposition of $LiNbO_3$ film on the diamond substrate without an $Al_2O_3$ buffer layer resulted in bubble formation and peeling off of the film at even 100 mTorr oxygen pressure due to oxidation of the diamond surface at high temperature in oxygen ambient (see Fig 2.2 (b)). Furthermore, the SEM image shows that interfacial reaction occurs, as shown in Fig. 2.2 (a), where a diffusion layer can be clearly observed in between the $LiNbO_3$ film and the diamond substrate. Therefore, it can be concluded that an $Al_2O_3$ buffer layer can effectively protect the diamond surface from oxidation at high temperature in oxygen ambient.

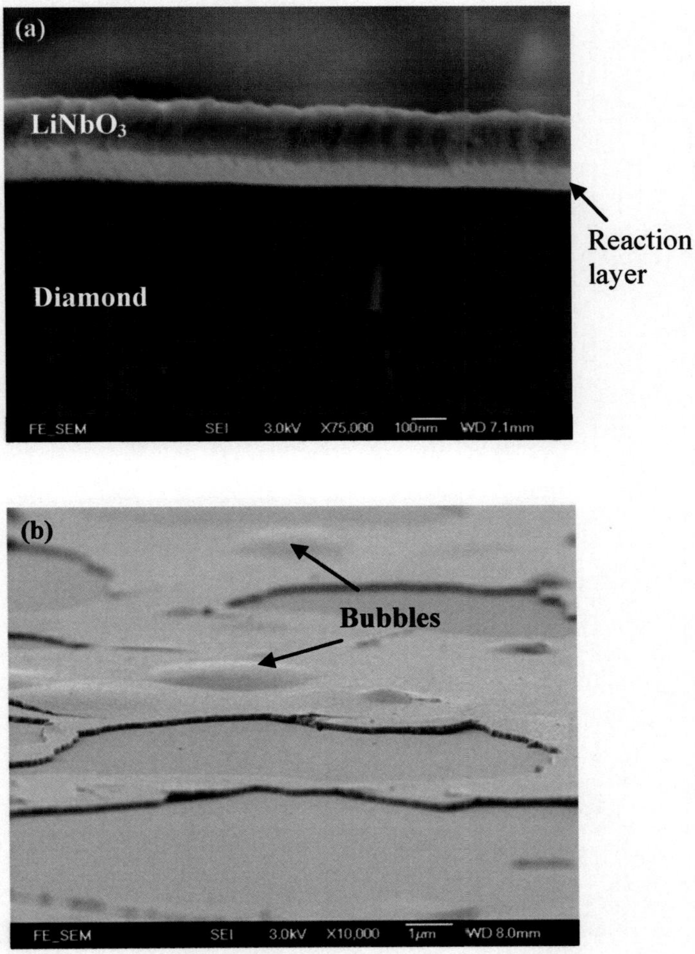


Figure 2.2. SEM images of $LiNbO_3$ film directly deposited on diamond substrate without $Al_2O_3$ buffer layer at 600 °C and 100 mTorr oxygen pressure: (a) cross-section, and (b) surface morphology.

Figures 2.3 and 2.4 are XRD patterns and AFM images, respectively, of the $LiNbO_3$ films on $Al_2O_3$ buffered diamond substrates deposited by the three different methods. XRD analysis (see Fig. 2.3 (a)) revealed that the $LiNbO_3$ film deposited at 500 mTorr oxygen pressure possesses strong c–axis preferred orientation, but the film surface is rough (r.m.s. roughness is 15 nm as shown in Fig. 2.4 (a)) due to the particulate problem in high-pressure PLD. The film deposited at 100 mTorr oxygen partial pressure, however, shows much lower r.m.s. roughness of 6 nm, but poor crystallinity (see Figs. 2.3 (b) and 2.4 (b)). In contrast, as shown in Figs. 2.3 (c) and 2.4 (c), the surface flatness of the two–step deposited $LiNbO_3$ film is similar to that deposited at 100 mTorr oxygen pressure and the film shows very strong (0006) $LiNbO_3$ diffraction peak, indicating highly c–oriented crystal orientation and improved crystallinity. It must be noted that the XRD spectra did not show any reflection from the $Al_2O_3$ buffer layers due to the fact that the $Al_2O_3$ has an amorphous structure.

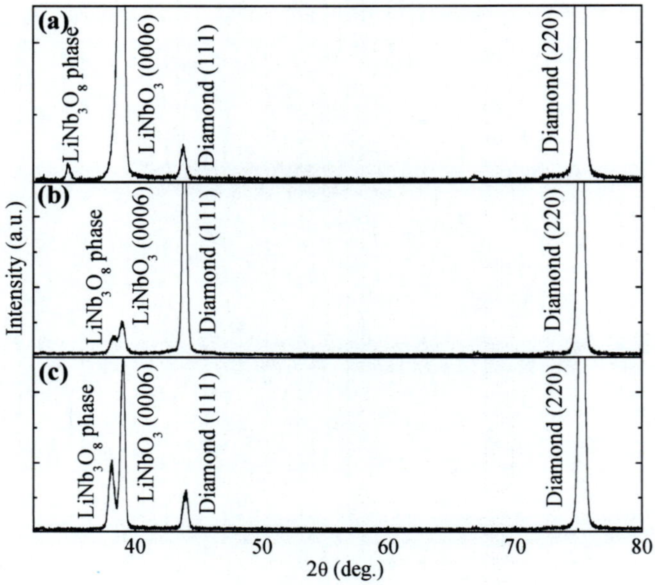


Figure 2.3. XRD patterns of the $LiNbO_3$ films grown on $Al_2O_3$ buffered diamond at 600 °C under different oxygen pressures: (a) 500 mTorr, (b) 100 mTorr, and (c) two–step method (500 mTorr for 15 s followed by 12 min at 100 mTorr).

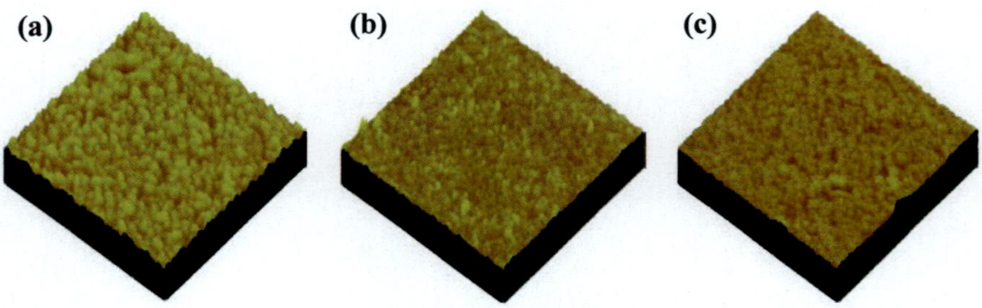


Figure 2.4. Three-dimensional AFM micrographs (5.0 μm × 5.0 μm) of the $LiNbO_3$ films grown on $Al_2O_3$ buffered diamond at 600 °C under different oxygen pressures: (a) 500 mTorr, (b) 100 mTorr, and (c) two–step method (500 mTorr for 15 s followed by 12 min at 100 mTorr).

Although the $LiNbO_3$ film shows a very strong c-oriented crystal orientation, the Li deficient phase $LiNb_3O_8$ is still present inside the film, which may deteriorate the performance of SAW device. A further improvement on optimizing the growth condition of

$LiNbO_3$ on $Al_2O_3$ buffered diamond, therefore, was carried out. By increasing the annealing oxygen pressure to 1 atm, the Li deficient phase $LiNb_3O_8$ disappears (see Figs. 2.5 (a) and (b)). The reason for the presence of $LiNb_3O_8$ phase inside the two-step deposited $LiNbO_3$ film is due to insufficient oxygen pressure during the annealing, resulting in vaporization of the Li atoms from the film. A higher oxygen pressure annealing environment is helpful to prevent Li evaporation, thus the stoichiometry of $LiNbO_3$ film can be maintained. It is worthy to note that the same growth condition for $LiNbO_3$ on sapphire substrate does not result in the formation of $LiNb_3O_8$ phase. This may be due to the fact that the $Al_2O_3$ buffer layer grown by the PLD method without introducing oxygen gas is oxygen deficient, thus the $Al_2O_3$ film should consume additional oxygen during the subsequent annealing of $LiNbO_3$ film. Therefore, higher oxygen pressure (1 atm) during annealing is able to compensate for the oxygen loss in the $Al_2O_3$ film and to eliminate the formation of $LiNb_3O_8$ phase.

At such a high temperature and high oxygen ambient deposition condition, $Al_2O_3$ buffer layer plays an important role in protecting polycrystalline diamond substrate from oxidation. In addition, the amorphous structure of $Al_2O_3$ can eliminate the heritage of polycrystalline structure of diamond and make the growth of c-oriented $LiNbO_3$ film possible. Such high-degree of c–axis orientation and low surface roughness features make the films suitable for SAW device application.

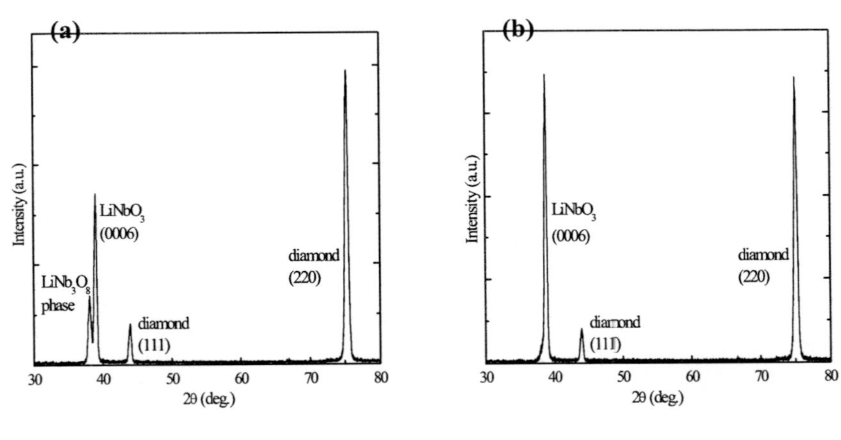


Figure 2.5. XRD patterns of $LiNbO_3/Al_2O_3$/diamond grown at 600 °C with the two-step method: (a) annealed at 11 Torr, and (b) annealed at 1 atm.

## 2.3. Growth Mechanism of $LiNbO_3$ on $Al_2O_3$ Buffer Layer

Gibbs free energy has been thought to play an important role in the nucleation stage which directly affects the growth orientation of the films. The total change of the Gibbs free energy $\Delta G$ in the formation of single crystal nucleus is dominated by the nucleus chemical potential $\Delta u$ and surface free energy $\sigma$ . It can be denoted as

$$\Delta G = -V\Delta u + S\sigma \quad (2)$$

where $V$ and $S$ are the nucleus volume and the nucleus surface area, and $\Delta u$ and $\sigma$ are the chemical potential per unit volume nucleus formation and the surface free energy per area surface formation, respectively [26]. For certain crystal material, the value of $\Delta u$ is defined, however, the value of $\sigma$ is different for different adjacent lattice planes. From the thermodynamics view point, the lattice planes grown with preference are those with lower surface free energy. The previous studies showed the random orientation of $LiNbO_3$ deposited on diamond substrate which may be attributed to an additional interfacial free energy resulted from the lattice mismatch. Therefore, $Al_2O_3$ amorphous buffer layer is believed to eliminate the interfacial free energy caused by the lattice mismatch. Since the $LiNbO_3$ is in hexagonal structure, the (0001) lattice plane has the largest atomic density and lowest surface free energy. Thus, it grows preferentially along the c-axis on amorphous substrates. By contrast, previous investigations [19-20] showed that random orientations of $LiNbO_3$ films were obtained from direct deposition on polycrystalline diamond substrate by different deposition techniques.

The merit of the two-step growth in enhancing the c-oriented $LiNbO_3$ crystalline may be understood by the model proposed by Lee *et al* [7]. Once a seed layer is successfully formed at high oxygen pressure, the high-energy ablated species and the substrate heating will provide enough energy to facilitate the diffusion of adatoms, and therefore the film can subsequently grow preferentially along the c-axis even at an unfavorable deposition ambient (low oxygen pressure).

## CONCLUSION

Pulsed-laser deposition of $LiNbO_3$ films on sapphire and diamond substrates has been carried out. The relationships between the film deposition conditions and the film structures, and their mechanisms, have been studied. The study of $LiNbO_3$ film deposition on different orientations of sapphire substrates by PLD under different conditions revealed that high oxygen pressure (500 mTorr) favors c-oriented growth and low oxygen pressure (100 mTorr) favors a-oriented growth of the $LiNbO_3$ films. Epitaxial $LiNbO_3$ film growth has been achieved by optimizing the growth condition. Since high oxygen pressure growth usually results in rough film surface, a novel two-step growth method (a seed layer deposition at 500 mTorr for 15 s followed by a deposition at 100 mTorr oxygen pressure) was developed to grow highly c-oriented and surface-flat $LiNbO_3$ films on sapphire substrates.

Three key factors have been optimized to obtain high-quality c-oriented $LiNbO_3$ films on polycrystalline diamond substrate which can be used for high-frequency SAW device fabrication. Firstly, an $Al_2O_3$ amorphous buffer layer grown by PLD was used as an oxygen barrier layer to prevent the diamond surface from oxidation under high temperature and oxygen pressure. Secondly, a two-step growth method was adopted in $LiNbO_3$ film growth on diamond substrate. Thirdly, high oxygen pressure (1 atm) annealing is found to be essential to eliminate Li deficient phase in the $LiNbO_3$ films. A growth mechanism in considering the surface free energy and Gibbs free energy has been proposed to interpret the orientation controllable growth of the $LiNbO_3$ films on sapphire substrates.

## REFERENCES


[1] D. Hofmann, Nonlinear optical devices. *http://fb6www.uni-paderborn.de/ag/ag-sol/research/qpm/qpmatch.htm,* Paderborn University, Paderborn.

[2] D. P. Morgan, "Surface acoustic wave devices and applications" *Ultransonics* 11, 121 (1973).

[3] Y. Shibata, K. Kaya and K. Akashi, *J. Appl. Phys.* 77, 1498 (1995).

[4] Y. Shibata, K. Kaya and K. Akashi, *Appl. Phys. Lett.* 61, 1000 (1992).

[5] Y. Shibata, Y. Kanno and K. Kaya, Jpn. *J. Appl. Phys.* 34, L320 (1995).

[6] M. Marsh, S. D. Harkness, and F. Qian, *Appl. Phys. Lett.* 62, 952 (1993).

[7] S. H. Lee, T. K. Song, T. W. Noh, and J. H. Lee, *Appl. Phys. Lett.* 67, 43 (1995).

[8] D. W. Kim, S. H. Lee and T. W. Noh, *Mater. Sci. Eng. B* 56, 251 (1998).

[9] G. Balestrino, S. Martellucci and P. G. Medaglia, *Appl. Phys. Lett.* 78, 1204 (2001).

[10] J. A. Chaos, V. Pruneri and J. Gonzalo, *J. Appl. Phys.* 88, 3768 (2000).

[11] X. C. Wang, J. H. He, *J. Mater. Sci. Lett.* 22, 225 (2003).

[12] X. C. Wang, Z. Z. Ye and J. H. He, *Int. J. Mod. Phys. B* 16, 4343 (2002).

[13] K. K. Wong, "Growth", *Properties of lithium niobate*, INSPEC, London, pp.38 (2002).

[14] S. H. Lee, T. W. Noh, J. H. Lee, *Appl. Phys. Lett.* 68, 472 (1996).

[15] Z. Z. Ye, J. H. He and L. F. Ye, *Mater. Lett.* 55, 265 (2002).

[16] M. Ishihara, T. Nakamura and F. Kokai, *Diamond Relat. Mater.* 11, 408 (2002).

[17] E. Dogheche, X. Lansiaux and D. Remiens, Jpn. *J. Appl. Phys.* 42, 573 (2003).

[18] M. Ishihara, T. Nakamura and F. Kokai, *Diam. Relat. Mater.* 12, 1809 (2003).

[19] T. C. Lee, J. T. Lee and M. A. Robert, *Appl. Phys. Lett.* 82, 191 (2003).

[20] E. Dogheche, V. Sadaune, X. Lansiaux, *Appl. Phys. Lett.* 81, 1329 (2002).

[21] Q. Wan, N. L. Zhang and L. W. Wang, *Thin Solid Films* 415, 64 (2002).

[22] W. S. Hu, Z. G. Liu and Z. C. Wu, *Appl. Surf. Sci.* 141, 197 (1999).

[23] T. A. Derouin, C. D. E. Lakeman and X. H. Wu, *J. Mater. Res.* 12, 1391 (1997).

[24] S. Kohiki, M. Arai. H. Yoshikawa, and S. Fukushima, *Phys. Rev. B* 57, 14572 (1998).

[25] L. Kasper, and S. Hüfner, *Phys. Lett. A* 81, 165 (1981).

[26] J. H. He and Z. Z. Ye, *Chinese Sci. Bull.* 48, 2290 (2003).

In: Trends in Crystal Growth Research
Editor: George V. Karas, pp. 23-44
ISBN 1-59454-541-3


*Chapter 2*

# EVALUATION AND GROWTH OF MICROCHIP LASER CRYSTAL

***Yidong Huang, Zundu Luo and Yanfu Lin***
Fujian Institute of Research on the Structure of Matter, Chinese Academy of Sciences,
Fuzhou, Fujian 350002, China

## ABSTRACT

Microchip laser is the most compact laser diode (LD) pumped solid-state laser with gain medium less than 1mm and can oscillate in a single longitudinal mode, so compared with the conventional solid-state laser, higher concentration of active ions is necessary for full absorption of the pumping power in such a thin microchip. However, the difficulty of crystal growth and the effect of fluorescence quenching increase more or less with the concentration of active ions doped in a crystal. In this chapter, taking into account the crystal growth and the fluorescence quenching, some laser crystals, such as $Nd^{3+}$ and $Yb^{3+}$ doped YAG, $KGd(WO_4)_2$, and $YVO_4$, and the stoichiometric neodymium crystals are evaluated as microchip laser media. The chemical structure reasons for weak concentration dependent fluorescence quenching in the borate laser crystals are discussed. The microchip laser crystal growth, including stoichiometric laser crystal $YbAl_3(BO_3)_4$, and crystals of $YAl_3(BO_3)_4$ and $GdAl_3(BO_3)_4$ doped with series of $Yb^{3+}$ concentrations, are analyzed. Two new ways for conquering the difficulty of microchip processing, the liquid phase epitaxy and the adoption of cleavage plates, are introduced.

## 1. INTRODUCTION

Microchip laser is a monolithic laser diode (LD) pumped solid-state device consisted of a slice of solid-state gain medium with thickness less than 1 mm and polished to be plane parallel, and the cavity mirrors deposited directly onto the medium faces [1]. Therefore, primarily, there are no internal surfaces to cause additional loss and no cavity alignment problems; furthermore, for such laser media, even low beam quality pump sources (like most of the LD) can be used, still achieving a good overlap of pump and laser beams. The simplicity and effectivity of these devices can lead to low unit cost manufacturing, making them highly competitive with other laser technology.

Particularly, the short laser cavity length of a microchip laser can provide a highly compact source of coherent and potentially single mode radiation. Thermal and gain effects can define a single transverse mode. In addition, the large free spectral range of the cavity can lead to only one cavity mode within the gain bandwidth, giving a single longitudinal mode laser output with high spatial ($TEM_{00}$ Gaussian mode) and spectral (single-longitudinal-mode, linewidth less than 5 kHz.) quality laser. So microchip laser is also a kind of optical transformer which transforms a poor quality LD beam to a diffraction limit $TEM_{00}$ and single frequency laser beam. There is still intense research into single-frequency lasers because there are numerous applications in which a multi-mode laser would be unacceptable or at least disadvantageous. These applications include high-resolution spectroscopy, where the narrow line width of a single-frequency laser is essential, and interferometry, which requires the long coherence length of a single-frequency laser. A single-frequency laser can also be used in coherent lidars, coherent optical communications, remote sensing, and many different industrial areas. In single-frequency lasers the beating between axial modes is eliminated, as there is only one axial mode, giving greater amplitude stability. This is especially advantageous for lasers used as a pump source for non-linear processes such as frequency doubling or pumping of an optical parameteric oscillator (OPO) if they ought to operate with small amplitude fluctuations. So it can be expected that the microchip lasers should open the domain of solid-state lasers to high volume and low cost markets [1-6].

In microchip lasers, because the thickness of medium is limited, it is necessary to have a high doping concentration of active ions in order to absorb adequate pump power. However, with the high doping concentration, the concentration dependent fluorescence quenching and the difficulty of crystal growth have to be taken into account. In this chapter, some $Nd^{3+}$ and $Yb^{3+}$-doped laser crystals are evaluated as microchip laser media. The chemical structure reasons for weak concentration dependent fluorescence quenching in the borate laser crystals are discussed. Some crystals grown in our lab, which may be suitable for microchip laser, are introduced. At last, the possible ways for conquering the difficulty of microchip processing are given out.

## 2. Evaluation of Microchip Laser Medium

### 2.1. $Nd^{3+}$-Doped Laser Crystal

Conventionally, several kinds of technique have been used practically to oscillate a laser in a single longitudinal mode, for example, mode discrimination by insertion of a Brewster plate inside the cavity and elimination of the standing-wave pattern by use of a pair of quarter-wave plates or a ring cavity configuration [7]. Microchip laser has a short cavity length and one of its advantages is the possible single longitudinal mode output. To obtain the single longitudinal mode output, the thickness of the microchip must be less than [7]

$$L_s = \frac{\lambda^2}{2n\Delta\lambda} \tag{1}$$

where $\lambda$ is the wavelength of microchip laser, $\Delta\lambda$ and $n$ are the laser emission gain bandwidth and the refraction index of the microchip, respectively, $c$ is the speed of light. From the data of $\Delta\lambda$ and $n$ for $Nd^{3+}$-doped YAG, $YVO_4$, and $LaSc_3(BO_3)_4$ crystals [8-12], the corresponding maximum thickness of microchip for single longitudinal mode output of the $Nd^{3+}$ laser around 1.06μm could be calculated and the values are listed in Table 1. Clearly, the thickness of the microchip must be less than 0.5mm. In other respect, the microchip laser is generally end-pumped by LDs and the single-pass absorption efficiency of the pumping power by the microchip with thickness of $L_s$ is [7]

$$\eta_a = 1 - \exp(-N\sigma_a L_s) \tag{2}$$

where $N$ is the active ion concentration in ions/cm$^3$, $\sigma_a$ is the absorption cross-section at the pump wavelength. To have an efficient laser operation, a high absorption efficiency is necessary. On the basis of the $\sigma_a$ at 807nm, which is the popular pump wavelength of $Nd^{3+}$ laser, for the three crystals listed in Table 1, the minimum concentration $N_{min}$ of $Nd^{3+}$ ions in the crystals for single longitudinal mode output with $\eta_a \geq 80\%$ could be estimated by

$$N_{\min} = \frac{2\ln(1-\eta_a)n\Delta\lambda}{\sigma_a \lambda^2} \tag{3}$$

and the results in at.% are listed in Table 1.

For a crystal grown with a dopant, its quality is generally more or less degraded with the increase of the dopant concentration. In Table 1, the reported highest $Nd^{3+}$ concentrations in crystals with good optical quality [8,13,14], $N_{limit}$, are listed. It is clear that due to the limitation of $Nd^{3+}$ concentration, part of the existing laser crystals, especially for those with wide gain band and small absorption cross-section, are unsuitable for microchip laser with single longitudinal mode output. For Nd:$YVO_4$ crystal, the $N_{min}$ is far lower than the $N_{limit}$, and the crystal also has the largest emission cross section $\sigma_e$ at 1.06μm among the crystals considered here (see Table 1), which will make its laser operate efficiently. Therefore, compared with the YAG and $LaSc_3(BO_3)_4$ crystals, the $YVO_4$ is more suitable for $Nd^{3+}$ microchip laser with single longitudinal mode output, although the rather poor thermo-mechanical properties of the crystal restrict severely the power level of laser output.

Also due to its large absorption cross-section, the Nd:$YVO_4$ crystal has been investigated as a solid state laser material for single longitudinal mode operation with short absorption depth [5,15]. Furthermore, single longitudinal mode operation of intracavity-doubled Nd:$YVO_4$ microchip laser has been realized [15,16]. The difficulty of the microchip fabrication can be avoided when single longitudinal mode output realized by this way; however, the requirement of high $Nd^{3+}$ concentration in crystals for the method is more indispensable than that for the microchip.

**Table 1. Properties of several $Nd^{3+}$-doped crystals for 1.06μm microchip laser with single longitudinal mode output**

| Host crystal | YAG | $YVO_4$ | $LaSc_3(BO_3)_4$ |
|---|---|---|---|
| $\sigma_e$ ($10^{-20}cm^2$) | 28 | 156 | 13 |
| $\Delta\lambda$ (nm) | 0.7 | 1.0 | 4.0 |
| $L_s$ (μm) | 462 | 274 | 77 |
| $\sigma_a$ for indirect pumping ($10^{-20}cm^2$) | 5.6 | 25 | 7.0 |
| $N_{min}$ for indirect pumping (at.%) | 4.5 | 1.9 | 65 |
| $\sigma_a$ for direct pumping ($10^{-20}cm^2$) | 1.2 | 15 | 1.2 |
| $N_{min}$ for direct pumping (at.%) | 21 | 3.1 | >100 |
| $N_{limit}$ in grown crystal (at.%) | 3.5 | 5 | 25 |

In convention, LD at wavelength of around 807nm is mostly used as a pumping source for $Nd^{3+}$ lasers due to the large absorption cross section of the transition $^4I_{9/2}\rightarrow{}^4F_{5/2}$ of the $Nd^{3+}$ ions. The ions excited to the $^4F_{5/2}$ level are populated the upper laser level $^4F_{3/2}$ by nonradiative process, so pumping at around 807nm for $Nd^{3+}$ laser is also called indirect pumping. Recently, the so-called direct pumping scheme, i.e. pumping the $Nd^{3+}$ ions into the $^4F_{3/2}$ level directly, has attracted much attention [14, 17-20] since the LD at wavelength of around 885nm corresponding to the transition of $^4I_{9/2}\rightarrow{}^4F_{3/2}$ is available. Compared with the indirect pumping, the direct pumping is regarded as a better way to get higher slope efficiency and lower pumping threshold since it reduces the quantum defect and so the thermal load [19-20].

However, as listed in Table 1, the absorption cross-section of $Nd^{3+}$ ion for the direct pumping [14,21,22] is markedly smaller than that for indirect pumping in the same crystal. In order to absorb adequate pumping power, higher $Nd^{3+}$ concentration, $N_{min}$, is necessary for keeping single longitudinal mode output. Clearly, some of the laser crystals, such as the Nd:$LaSc_3(BO_3)_4$, are not suitable for direct pumping while keeping single longitudinal mode output even the problem of crystal growth caused by the high dopant concentration is not taken into account. For the Nd:YAG crystal, the $Nd^{3+}$ concentration higher than 20at.% is also impossible for crystal growth. To our knowledge, for the crystal, the highest $Nd^{3+}$ concentration reported is 3~4at.% [11]. Even for $Nd^{3+}$:YAG ceramic, the highest $Nd^{3+}$ concentration is only 9at.% for keeping the spectral parameters and the laser performances almost the same as those of single crystal [23]. Therefore, even using $Nd^{3+}$:YAG ceramic as a microchip, the efficient single longitudinal mode output can not be realized by direct pumping. Besides the difficulty of crystal growth, high $Nd^{3+}$ concentration will generally bring some other problems such as the concentration dependent fluorescence quenching and energy transfer between the $Nd^{3+}$ ions, etc. All of these will cause a low efficiency of laser operation. The detailed analysis on the laser performances of $Nd^{3+}$-activated crystals under indirect and direct pumping can be found in Ref. [24].

Although high $Nd^{3+}$ concentration enough for microchip laser with single longitudinal mode output under indirect and direct pumping is possible for the Nd:$YVO_4$ crystal, the energy transfer processes in the Nd:$YVO_4$ crystal are much stronger than those in the Nd:YAG and a strong reduction of emission quantum efficiency at high $Nd^{3+}$ concentration could take place [25]. This would limit a high efficiency of single longitudinal mode laser emission from the high $Nd^{3+}$-doped $YVO_4$ crystal under direct pumping.

## 2.2. Stoichiometric Neodymium Crystal

Considering a high active ion concentration is necessary for microchip laser, one special class of laser crystals, in which $Nd^{3+}$ ion is a stoichiometric constituent, i.e. 100at.% concentration, rather than a dopant [26,27], has attracted much attention as potential microchip laser media [28-30]. In a stoichiometric laser crystal, it is also called self-activated crystal, the $Nd^{3+}$ concentration could be higher than $5\times10^{21}cm^{-3}$ [26], which is about 40 times higher than in Nd:$YVO_4$ with 1at.% doping. As a result, the end-pumping power can be absorbed fully over short distances, typically about 100μm. Furthermore, since $Nd^{3+}$ ions are not incorporated as inhomogeneities in crystal lattice, the optical quality of these crystals is not affected by the high $Nd^{3+}$ concentration.

The stoichiometric neodymium crystals have been the subject of considerable investigation since 30 years ago [26], when the technique of microchip laser was not proposed, based on their potential application in efficient miniature solid-state lasers. Some of the main stoichiometric neodymium crystals with their $Nd^{3+}$ concentration $N_s$ are listed in Table 2 [26,27,31]. Among them, the microchip lasers have been realized in the $NdAl_3(BO_3)_4$ [29,30] and $LiNd(PO_3)_4$ [28] crystals. For the $KNd(PO_3)_4$ [32,33], $NdP_5O_{14}$ [34], and $K_5Nd(MoO_4)_4$ [31,35] crystals, other types of solid-state laser have been obtained.

In the selection of a stoichiometric neodymium crystal as a laser medium, weak concentration dependent fluorescence quenching is the first consideration. In Table 2, the fluorescence lifetimes of the upper laser level $^4F_{3/2}$ for the crystals with low $Nd^{3+}$ concentration of 1.0at.% ($\tau_0$) and the stoichiometric neodymium crystals ($\tau_1$) are compared. The values of fluorescence quenching ratio $q$ defined by [32]

$$q = \tau_0 / \tau_1 \tag{4}$$

are also listed in Table 2. It can be found that the values of the quenching ratio are between 2.41 and 3.07 for all the stoichiometric neodymium laser crystals. Thus, the quenching rate for the stoichiometric crystals, $1/\tau_1$-$1/\tau_0$, is approximately proportional to the predominantly radiative decay rate, $1/\tau_0$. In addition, the observed dependence of fluorescence lifetime on $Nd^{3+}$ concentration shows that the quenching rate is roughly proportional to concentration when the concentration higher than 10at.%. In contrast, the quenching rate in Nd:YAG increases as the square of the concentration, which is characteristic of dipole-dipole interactions between $Nd^{3+}$ ions [26]. The proportionality of the quenching rate to both $1/\tau_0$ and concentration may originate from quenching by exchange interactions between $Nd^{3+}$ ions [26].

It is generally accepted [26] that the fluorescence quenching is relatively weak in crystals with large Nd-Nd spacing and anion-cation-anion links between $Nd^{3+}$ ions. Auzel proposed a criterion [36] that if the crystal-field strength parameter of a stoichiometric neodymium crystal satisfies the relation

$$N_v = \left( \sum_{k,q} B_{k,q}^{\ 2} \right)^{1/2} \leq 1800\text{cm}^{-1} \tag{5}$$

the fluorescence quenching will be weak enough and make the crystal as a potential laser medium. For some of the stoichiometric neodymium crystal, the values of $N_v$ could be calculated from the crystal-field parameters $B_{k,q}$ available in the literature. For the $KNd(PO_3)_4$ and $K_5Nd(MoO_4)_4$ crystals, although the values of $B_{k,q}$ are unavailable in the literature at the present time, the crystal-field splitting of the manifolds for $Nd^{3+}$ ions has been determined [31,37]. Then the $N_v$ could be estimated by [38]

$$\frac{N_{v,A}}{N_{v,B}} = \frac{\Delta E_A}{\Delta E_B} \tag{6}$$

where subscripts $A$ and $B$ represent two different $Nd^{3+}$-doped crystals, $\Delta E$ is the maximum splitting of a manifold for the $Nd^{3+}$ ions in the crystals. From the maximum splitting of manifolds [39] and $N_v$ calculated from $B_{k,q}$ for the $NdAl_3(BO_3)_4$ crystal, the values of $N_v$ for $KNd(PO_3)_4$ and $K_5Nd(MoO_4)_4$ crystals could be obtained. All the results are listed in Table 2. It could be found that, except $K_5Nd(MoO_4)_4$, all of the stoichiometric neodymium crystals in which laser operation has been realized obey the relation of Eq. (5). This criterion implies that the melting point of the stoichiometric neodymium laser crystals should not exceed 1200°C [27]. In other words, the structures of these crystals must not be as compact as the conventional doped laser crystals.

However, a laser crystal with such a structure should also be as compact as possible for meeting the requirements of laser devices. In order to take these two contradictory requirements simultaneously into account, it is necessary to investigate the relationship between the chemical composition and structural characteristics on the one hand and effect of fluorescence quenching on the other.

Obviously, different kinds of anion-cation-anion links have different degrees of inhibition effect on the fluorescence quenching besides the geometry factor. Here, a parameter defined by [27]

$$L = R^- / R^+ \cdot \Delta x \tag{7}$$

is proposed to describe the effect of the link on the inhibition of fluorescence quenching, where $\Delta x$ is the difference in electronegativities between the anion and the cation within the links, and $R^+$ and $R^-$ are the ionic radii of the cation and the anion respectively. The values of $L$ for the stoichiometric neodymium laser crystals were calculated and are listed in Table 2. Compared with the $PO_n$, $WO_n$, and $MoO_n$ groups, the $BO_n$ group has the largest value 4.3 of $L$. In experiments, the $NdAl_3(BO_3)_4$ crystal has a low fluorescence quenching ratio, yet has the highest $Nd^{3+}$ concentration among all the stoichiometric neodymium laser crystals. It can be seen from Table 2 that a larger value of $L$ invariably corresponds to a larger value of $Nd^{3+}$ concentration per unit quenching ratio

$$N_q = N_s / q \tag{8}$$

From the analysis given above, it follows that it is possible for a crystal with the largest value of $L$ to possess a considerable compact structure and at the same time a large number of luminescence centers per unit volume, i.e. these two mutually contradictory requirements can be simultaneously fulfilled. The results of microchip laser experiments also demonstrate the validity of parameter $L$ as a criterion for judging a stoichiometric neodymium crystal as a microchip laser medium. For example, with the same thickness of 0.3mm and similar LD pumping power of 300mW, the pump-to-laser efficiency of the $NdAl_3(BO_3)_4$ crystal is about 30% [29] while that of the $LiNd(PO_3)_4$ crystal is less than 20% [28].

In spite of the good properties of the $NdAl_3(BO_3)_4$ crystal, its application in high-power diode pumped microchip lasers is limited because of the strong pump-induced crystal heating caused by the high nonradiative de-excitation rates of $Nd^{3+}$ ions in the crystal [40]. Therefore, the reduction of the thermal load in a $NdAl_3(BO_3)_4$ laser is a fundamental problem which should be addressed before the development of high-power microchip lasers based on this material. As mentioned above, the direct pumping is an efficient way to reduce the thermal load especially for the stoichiometric neodymium laser crystals. Its high $Nd^{3+}$ concentration could compensate for the small absorption cross-section at the direct pumping wavelength. The reduction in the pump-induced thermal load can improve the laser performance of $NdAl_3(BO_3)_4$ microchip laser by minimizing the thermal effects. Laser experiments have shown that the slope efficiency is improved more than 20% (from 50% up to 70%) when the pump wavelength is changed from 808nm up to 882nm [30]. As a consequence, direct pumping is a powerful tool to improve the laser performance of stoichiometric neodymium microchip lasers.

## 2.3. $Yb^{3+}$-Doped Laser Crystal

Since high power InGaAs LD was used as a pump source of the solid state lasers, the $Yb^{3+}$-doped laser materials have received much attention because they can be efficiently pumped by the InGaAs LD. Compared with their $Nd^{3+}$-doped counterparts, the $Yb^{3+}$-doped materials have some advantages [41,42]. The radius of the $Yb^{3+}$ ion is smaller than that of $Nd^{3+}$, which makes it easier to be doped into the host media and the doping concentration can be very high, for example, the YAG, $KY(WO_4)_2$, $YAl_3(BO_3)_4$, and $GdAl_3(BO_3)_4$ with a series of $Yb^{3+}$ concentrations to as high as 100at.%, i.e. the stoichiometric ytterbium crystal, have been successfully grown [43-47]. Because the $Yb^{3+}$ ion has only two electronic states separated by approximately 10000 $cm^{-1}$, i.e. the $^2F_{7/2}$ ground state and the $^2F_{5/2}$ excited state, some deleterious effects such as excited state absorption, cross relaxation, and up-conversion are absent. The absence of additional 4f level in the ions also inhibits the concentration dependent fluorescence quenching, so laser crystals with high $Yb^{3+}$ concentration have weaker fluorescence quenching compared with the $Nd^{3+}$ crystals. The $Yb^{3+}$ ion also has a low quantum defect and so thermal load as a result of the small energy mismatch between the pump and laser wavelengths. The $Yb^{3+}$-doped materials have longer fluorescence lifetime of the upper laser level than those of their $Nd^{3+}$ counterparts, which is advantage for energy storage. Furthermore, the $Yb^{3+}$-doped materials exhibit broad absorption bands, so there is no need to precisely control the temperature of LD. One disadvantage of $Yb^{3+}$-doped laser materials is that they are quasi-three-level systems, which increases their pump thresholds, and make their laser performance strongly depend on the temperature.

Nowadays, many $Yb^{3+}$-doped laser crystals have been reported. Some spectroscopic parameters related to the microchip laser properties for several $Yb^{3+}$-doped crystals are listed in Table 3 [41,42,48-51]. Comparing the absorption and emission cross-sections of $Nd^{3+}$ listed in Table 1 and those of $Yb^{3+}$ listed in Table 3, respectively, it can be found that the absorption cross-sections of $Yb^{3+}$ are several times smaller than those of $Nd^{3+}$ for indirect pumping but similar to those for direct pumping while the emission cross-sections of $Yb^{3+}$ are about one order less than those of $Nd^{3+}$. On the other hand, the emission gain bandwidths of $Yb^{3+}$ are about one order broader than those of $Nd^{3+}$. As a result, the efficient single longitudinal mode output from the $Yb^{3+}$ microchip laser is impossible by the way of just using thin microchip medium. For example, the Yb:$KGd(WO_4)_2$ crystal has the largest absorption cross-section of $12\times10^{-20}cm^2$ at 981nm with the general emission gain bandwidth of 25nm around 1025nm among the $Yb^{3+}$-doped laser crystals. Estimated by Eq. (1), the thickness of the crystal for a single longitudinal mode laser output must be less than 10.5μm. The maximum $Yb^{3+}$ concentration in the crystal, i. e. the stoichiometric $KYb(WO_4)_2$ crystal, is $6.4\times10^{21}cm^{-3}$ [45]; therefore, calculated by Eq. (2), the possible highest single-pass absorption efficiency of the pumping power by the Yb:$KGd(WO_4)_2$ microchip with single longitudinal mode laser output is less than 55%.

However, the broad emission gain band makes the $Yb^{3+}$-doped crystal very suitable for tunable [35] and ultra-fast [6,52] microchip lasers. Simple and compact sources of sub-nanosecond pulses with high pulse energies and peak powers, together with a diffraction-limited output beam, are of interest for applications in the fields of range finding and non-linear frequency conversion. Diode-pumped passively Q-switched microchip lasers are ideal sources for these requirements. The short laser cavity lead to short pulses because the pulse width is directly proportional to the cavity round-trip time [10]. Especially, the $Yb^{3+}$-doped $Sr_3Y(BO_3)_3$ [49], $Ca_4GdO(BO_3)_3$ [53], and $Y_2SiO_5$ [48] crystals, which have emission gain bandwidth broader than 40nm, are the potential media for the tunable and ultra-fast microchip lasers.

Though $Yb^{3+}$-doped media have no concentration dependent fluorescence quenching by conventional processes due to the simple energy level structure of $Yb^{3+}$ ions, the quenching has been observed in experiments [43,54,55]. The unavoidable fluorescence quenching may be caused by a special mechanism so-called self-generated multiphonon-assisted defect center with increased electron-phonon coupling that cannot be suppressed whatever the host purity [56]. In $Yb^{3+}$ microchip lasers, because the thickness of laser crystal is limited and the absorption cross-section is relatively small, it is necessary to have a high doping concentration of active ions in order to get adequate pump power. Therefore, the fluorescence quenching effect must be considered. Also due to the high doping concentration, the thermal effect becomes noteworthy [57]. The lower level of the $Yb^{3+}$ laser is in the ground manifold, so the Boltzmann distribution in the microchip, caused by the temperature distribution, must be considered as well.

**Table 2. Properties of stoichiometric neodymium crystals ($x$=1 in the molecular formulas)**

| Crystal | $Nd_xGd_{1-x}Al_3(BO_3)_4$ | $LiNd_xLa_{1-x}(PO_3)_4$ | $KNd_xGd_{1-x}(PO_3)_4$ | $Nd_xLa_{1-x}P_5O_{14}$ | $Na_5Nd_xLa_{1-x}(WO_4)_4$ | $K_5Nd(MoO_4)_4$ |
|---|---|---|---|---|---|---|
| $N_s$ ($10^{21}$cm$^{-3}$) | 5.4 | 4.4 | 4.1 | 3.9 | 2.6 | 2.4 |
| $\tau_0$ (μs) | 50 | 325 | 275 | 320 | 220 | 215 |
| $\tau_1$ (μs) | 19 | 135 | 100 | 115 | 85 | 70 |
| $q$ | 2.63 | 2.41 | 2.75 | 2.78 | 2.59 | 3.07 |
| $N_v$ (cm$^{-1}$) | 1630 | 1474 | 1530 | 1406 | - | 2500 |
| $L$ | 4.3 | 2.7 | 2.7 | 2.7 | 1.5 | 1.5 |
| $N_q$ ($10^{21}$cm$^{-3}$) | 2.05 | 1.83 | 1.49 | 1.40 | 1.00 | 0.77 |

**Table 3. Properties of several $Yb^{3+}$-doped crystals**

| Crystal | YAG | $KGd(WO_4)_2$ | $Sr_5(PO_4)_3F$ | $Ca_4GdO(BO_3)_3$ | $YAl_3(BO_3)_4$ | $Y_2SiO_5$ | $Sr_3Y(BO_3)_3$ |
|---|---|---|---|---|---|---|---|
| $\sigma_e$ ($10^{-20}$cm$^2$) | 2.3 | 2.7 | 7.3 | 0.5 | 0.8 | 0.29 | 0.2 |
| $\Delta\lambda$ (nm) | 9 | 25 | 4 | 44 | 20 | 40 | 60 |
| $\sigma_a$ ($10^{-20}$cm$^2$) | 0.8 | 12 | 8.6 | 1.1 | 3.4 | 0.47 | 1.0 |

Due to the quasi-three-level in nature, the fluorescence spectrum and lifetime of $Yb^{3+}$ ions are strongly affected by the effect of radiation trapping, even when $Yb^{3+}$ concentration is as low as $3.6\times10^{19}cm^{-3}$ [43,58-60]. In order to obtain the intrinsic fluorescence spectrum and lifetime of $Yb^{3+}$-doped laser materials to evaluate their laser properties, especially for the materials with high $Yb^{3+}$ doping such as the microchip laser media, the effect of radiation trapping has been widely investigated and various methods have been proposed to limit it [58,60-62].

On the basis of a quasi-three-level system and taking into account the fluorescence quenching and the temperature distribution, a theoretical model of fundamental continuous-wave $Yb^{3+}$ microchip has been proposed [63,64]. In the model, the relationship between the output laser power as well as the pump threshold and the doping concentration along with the crystal thickness is investigated. As a result, the scheme of the optimization of $Yb^{3+}$ microchip laser is given out.

# 3. GROWTH OF MICROCHIP LASER CRYSTAL

## 3.1. Growth of $Yb^{3+}$-Doped Aluminium Borate Crystals

As mentioned in the last section, the borate crystals generally have a larger value of active ion concentration per unit quenching ratio. Furthermore, the zero static dipole moment of the $BO_3$ groups reduces drastically the internal vibration contribution to the rates of nonradiative transition and energy transfer, even though the frequency of some $BO_3$ vibration modes may be very high [65]. On the basis of successful growth of $NdAl_3(BO_3)_4$, $Nd{:}YAl_3(BO_3)_4$, and $Nd{:}GdAl_3(BO_3)_4$ crystals [27,66,67], and taking into account the microchip lasers realized in the low doping $Yb{:}YAl_3(BO_3)_4$ crystals [68,69], the aluminium borate crystals with a series of $Yb^{3+}$ concentrations, $Yb_xY_{1-x}Al_3(BO_3)_4$ ($x$=0.035, 0.065, 0.100, 0.173, 0.300, 0.407, 0.500, 0.655, and 0.750), $Yb_xGd_{1-x}Al_3(BO_3)_4$ ($x$=0.125, 0.330, 0.509, and 0.740), and $YbAl_3(BO_3)_4$, have been grown in our lab.

The $YAl_3(BO_3)_4$ crystals melt incongruently at 1290°C and can only be grown by the flux method [70]. The $Yb{:}YAl_3(BO_3)_4$ crystals have been grown by means of the top seeded solution growth (TSSG) method using $K_2Mo_3O_{10}$-$B_2O_3$ as solvents [71].

In our experiments, $Yb{:}YAl_3(BO_3)_4$ crystals were also grown by the TSSG method. The raw materials of $Al_2O_3$, $Y_2O_3$, $Yb_2O_3$, $H_3BO_3$, $K_2CO_3$ and $MoO_3$ were weighted according to the phase diagram in Ref. [72]. All the weighed high-purity chemicals were placed into a 60mm diameter and 60mm high platinum crucible and heated in a vertical tubular muffle furnace with a chrome-aluminium-molybdenum wire as heating element. The chemical reactions taking place in the platinum crucible is as follows:

$$xYb_2O_3+(1\text{-}x)Y_2O_3+3Al_2O_3+8H_3BO_3\rightarrow Yb_xY_{(1-x)}Al_3(BO_3)_4+12H_2O\uparrow$$
$$K_2CO_3+3MoO_3\rightarrow K_2Mo_3O_{10}+CO_2\uparrow$$

This mixture of the raw materials was heated to a temperature 100°C higher than the saturation temperature, which was determined by repeated seeding, and had been kept at this temperature for 48h to homogenize the melt. For raw materials with different $Yb^{3+}$

concentrations, the saturation temperature had a small change. After homogenization of the solution, the temperature was quickly decreased to 20°C higher than the saturation temperature. A seed crystal of $Yb:YAl_3(BO_3)_4$ oriented along c-axis was preheated and had been immerged in the solution for thirty minutes. Then the temperature of the melt was quickly decreased to the saturation temperature. In the process of growth, the crystal was rotated at 4.5rpm and cooled at 1.5°C/day. After a period of 30 to 40 days, a single crystal was drawn above the surface of the solution and cooled down to room temperature at a rate of 40°C/h.

Using the growth method described above, $Yb_xY_{1-x}Al_3(BO_3)_4$ crystals in a wide range of $x$ ($x$=0.035, 0.065, 0.100, 0.173, 0.300, 0.407, 0.500, 0.655, and 0.750) have been obtained. It is found that the high quality and large single crystals with $Yb^{3+}$ concentrations around 50at.% are more difficult to be grown.

The $Yb^{3+}$ concentrations in some of the grown crystals were measured by a Shimadzu 8705QH$_2$ electron probe X-ray microanalyser (EPMA) and an Oxford Instruments INCA energy dispersive spectrometer (EDS). The segregation coefficient $K$ of $Yb^{3+}$ in the $YAl_3(BO_3)_4$ crystal can be expressed as [73]

$$K = N_{cry} / N_{diss} \tag{9}$$

where $N_{cry}$ and $N_{diss}$ are the $Yb^{3+}$ concentrations in grown crystals and borate crystalline substances of the melt, respectively. The results of the microprobe analysis are listed in Table 4 [46,73]. It can be seen from the table that the segregation coefficients of the $Yb:YAl_3(BO_3)_4$ crystals are approximate to unit for different $Yb^{3+}$ concentrations.

**Table 4. $Yb^{3+}$ concentrations and segregation coefficients in $Yb:YAl_3(BO)_3$ crystals**

| Composition in melt | Composition in crystal | Segregation coefficient |
|---|---|---|
| $Yb_{0.005}Y_{0.995}Al_3(BO_3)_4$ | $Yb_{0.0047}Y_{0.9953}Al_3(BO_3)_4$ | 0.94 |
| $Yb_{0.025}Y_{0.975}Al_3(BO_3)_4$ | $Yb_{0.026}Y_{0.974}Al_3(BO_3)_4$ | 1.04 |
| $Yb_{0.05}Y_{0.95}Al_3(BO_3)_4$ | $Yb_{0.046}Y_{0.944}Al_3(BO_3)_4$ | 0.92 |
| $Yb_{0.172}Y_{0.828}Al_3(BO_3)_4$ | $Yb_{0.173}Y_{0.827}Al_3(BO_3)_4$ | 1.01 |
| $Yb_{0.384}Y_{0.616}Al_3(BO_3)_4$ | $Yb_{0.407}Y_{0.593}Al_3(BO_3)_4$ | 1.06 |
| $Yb_{0.652}Y_{0.348}Al_3(BO_3)_4$ | $Yb_{0.655}Y_{0.345}Al_3(BO_3)_4$ | 1.00 |

The grown $Yb_xY_{1-x}Al_3(BO_3)_4$ crystals with $x$=0.035, 0.100, 0.300, 0.500, 0.655, and 0.750 are shown in Fig. 1, all the crystals shown here have a size larger than 15×10×10mm$^3$. Although the centers of some grown crystals were opaque, samples of good optical quality with sizes up to 3.0×3.0×1.0mm$^3$ could be cut from the outer part of the crystals. The size is sufficient for spectral measurement and microchip laser applications.

Figure 1. Photograph of the as-grown $Yb_xY_{1-x}Al_3(BO_3)_4$ crystals.

By a similar way, $Yb_xGd_{1-x}Al_3(BO_3)_4$ crystals with $x$=0.125, 0.330, 0.509, and 0.740 have also been obtained. Different from the $Yb{:}YAl_3(BO_3)_4$ crystals, large single $Yb{:}GdAl_3(BO_3)_4$ crystals with high optical quality are difficult to be grown with the increase of $Yb^{3+}$ concentration. The $Yb^{3+}$ concentrations in the $Yb{:}GdAl_3(BO_3)_4$ crystals were measured by inductively coupled plasma atomic emission spectrometry (ICP-AES). The grown $Yb_xGd_{1-x}Al_3(BO_3)_4$ crystals with $x$=0.125, 0.330, and 0.509 are shown in Fig. 2, all the crystals shown here have a size larger than 20×10×10mm$^3$.

The segregation coefficient $K$ of $Yb^{3+}$ in the $GdAl_3(BO_3)_4$ crystals can be estimated by Eq. (9). From the values of $N_{cry}$, $N_{diss}$, and $K$ listed in Table 5 [47], it can be seen that the segregation coefficients of the $Yb{:}GdAl_3(BO_3)_4$ crystals change from 1.25 to less than 1.0 with the increase of $Yb^{3+}$ concentration. This is different from $Yb{:}YAl_3(BO_3)_4$ and may be the main reason for the optical quality of $Yb{:}GdAl_3(BO_3)_4$ crystals lower than that of $Yb{:}YAl_3(BO_3)_4$ under the similar growing conditions.

Figure 2. Photograph of the as-grown $Yb_xGd_{1-x}Al_3(BO_3)_4$ and $YbAl_3(BO_3)_4$crystals.

**Table 5. $Yb^{3+}$ concentrations and segregation coefficients in Yb:$GdAl_3(BO)_3$ crystals**

| Composition in melt | Composition in crystal | Segregation coefficient |
|---|---|---|
| $Yb_{0.10}Gd_{0.90}Al_3(BO_3)_4$ | $Yb_{0.125}Gd_{0.875}Al_3(BO_3)_4$ | 1.25 |
| $Yb_{0.30}Gd_{0.70}Al_3(BO_3)_4$ | $Yb_{0.33}Gd_{0.67}Al_3(BO_3)_4$ | 1.10 |
| $Yb_{0.50}Gd_{0.50}Al_3(BO_3)_4$ | $Yb_{0.509}Gd_{0.491}Al_3(BO_3)_4$ | 1.02 |
| $Yb_{0.75}Gd_{0.25}Al_3(BO_3)_4$ | $Yb_{0.74}Gd_{0.26}Al_3(BO_3)_4$ | 0.99 |

The X-ray powder diffraction (XRD) method was used to check the lattice structure of the grown crystals. Intensities of the diffraction peaks were recorded in a range of 10-80° (2θ) with a step of 0.02° and a scan speed of 5°/min. For example, the XRD patterns of the 65.5at.% $Yb^{3+}$-doped $YAl_3(BO_3)_4$ and the 50.9at.% $Yb^{3+}$-doped $GdAl_3(BO_3)_4$ are shown in Figs. 3 and 4, respectively. From the figures, it can be found that the diffraction patterns of both crystals are not only similar to each other but also similar to those of pure $YAl_3(BO_3)_4$, $GdAl_3(BO_3)_4$, and $YbAl_3(BO_3)_4$ crystals, and no additional peaks. It means that the high $Yb^{3+}$-doped crystals have the same structure as the pure crystals.

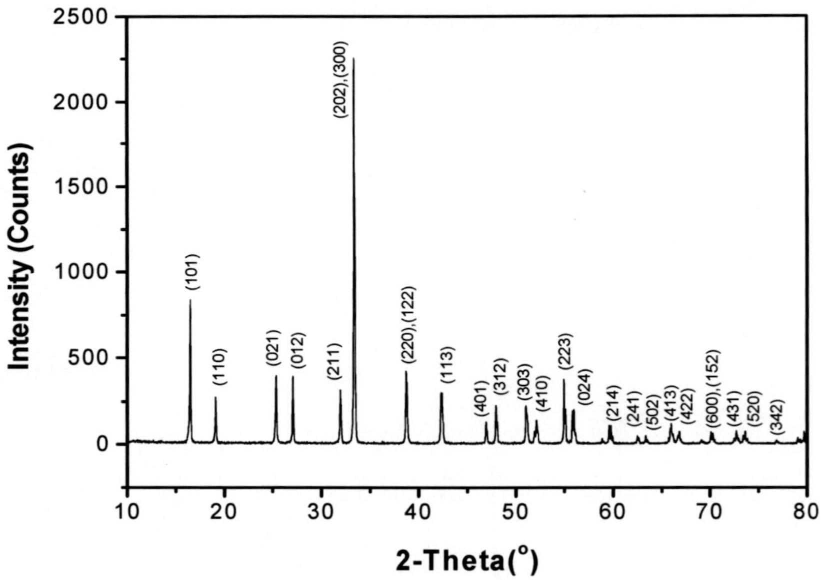


Figure 3. XRD pattern of 65.5at.% Yb:$YAl_3(BO_3)_4$ crystal at room temperature.

The stoichiometric $YbAl_3(BO_3)_4$ crystal has also been grown by the same way in our lab [74]. The grown $YbAl_3(BO_3)_4$ crystal with a size up to 20mm×10mm×10mm is also shown in Fig. 2. The crystal can be grown more easily than the $Yb^{3+}$-doped $YAl_3(BO_3)_4$ and $GdAl_3(BO_3)_4$ crystals.

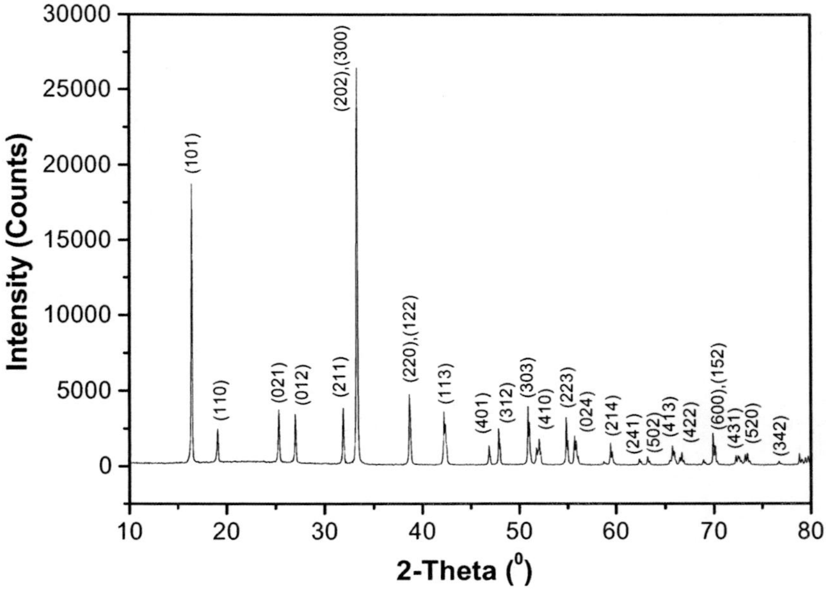


Figure 4. XRD pattern of 50.9at.% Yb:$GdAl_3(BO_3)_4$ crystal at room temperature.

The crystal structure of $YbAl_3(BO_3)_4$ was determined by the X-ray diffraction analysis [75]. The compound belongs to the trigonal system with the space group R32. The cell parameters are $a$=0.92512nm, $c$=0.71893nm, $Z$=3, $V$=0.53284nm$^3$ and the density of the crystal is $D_c$=4.574g/cm$^3$. The crystal has the same structure and similar cell parameters as $YAl_3(BO_3)_4$ ($a$=0.9293nm, $c$=0.7245nm) and $GdAl_3(BO_3)_4$ ($a$=0.9251nm, $c$=0.7189nm) crystals [76].

The spectroscopic parameters of the low and high $Yb^{3+}$-doped $YAl_3(BO_3)_4$ and $GdAl_3(BO_3)_4$ crystals are compared in Table 6 [46,47,77]. To restrain the radiation trapping effect, the powder method [58] was adopted in the measurements of the fluorescence spectra and lifetimes. All the samples were in type of fine powder with particle diameter of about 150μm and immersed into monochlorobenzene fluid (refractive index $n$=1.5240) contained by glass cuvettes. The purpose of the refractive index matching fluid used here was to reduce the total internal reflection of the fluorescence in the particles and then the probability of re-absorption. From Table 6, it can be found that the fluorescence lifetime $\tau_f$ of crystals with high $Yb^{3+}$ concentration are shorter than that with low $Yb^{3+}$ concentration. It is clear that although there is no excited absorption or cross-relaxation in the $Yb^{3+}$-doped crystals, the concentration dependent fluorescence quenching still exists. The abnormally long fluorescence lifetime of the 5.6at.% Yb:$YAl_3(BO_3)_4$ [77] may be a result of the radiation trapping effect.

**Table 6. Spectroscopic parameters of the low and high $Yb^{3+}$-doped $YAl_3(BO_3)_4$ and $GdAl_3(BO_3)_4$ crystals**

| Parameters | $YAl_3(BO_3)_4$ | | $GdAl_3(BO_3)_4$ | |
|---|---|---|---|---|
| $N_{cry}$ (at.%) | 5.6 | 65.5 | 12.5 | 50.9 |
| Main absorption peak (nm) (E⊥C) | 937, 975, 981 | 938, 976 | 936, 978 | 939, 976 |
| FWHM (nm) | 20 (975nm) | 22 (976nm) | 22 (978nm) | 22 (976nm) |
| $\sigma_a$ ($10^{-20}$cm$^2$) (E⊥C) | 3.4 (975nm) | 2.6 (976nm) | 2.5 (978nm) | 2.5 (976nm) |
| Main fluorescence peak (nm) | 998, 1040 | 999, 1038 | 991, 1040 | 991, 1040 |
| FWHM (nm) | - | 31 (1038nm) | 27 (1040nm) | 31 (1040nm) |
| $\sigma_e$ ($10^{-20}$cm$^2$) (E⊥C) | 0.8 | 1.2 (1038nm) | 0.7 (1040nm) | 0.7 (1040nm) |
| $\tau_f$ (μs) | 680 | 23 | 298 | 44.8 |
| $M$ ($10^{-23}$ cm·s) | 5.18 | 2.35 | 3.66 | 2.24 |

It is well known that the threshold of a continuous wave laser is inversely proportional to the merit factor [78]

$$M = \sigma_a \sigma_e \tau_f N_{cry} \tag{10}$$

A crystal with a large $M$ factor may be a good candidate for solid-state laser. From the values of $\sigma_a$, $\sigma_e$, and $\tau_f$ listed in Table 6, the $M$ factors for $Yb^{3+}$-doped $YAl_3(BO_3)_4$ and $GdAl_3(BO_3)_4$ crystals with different $Yb^{3+}$ concentrations were calculated and are also listed in Table 6. The results indicate that, although the fluorescence concentration quenching may cause the threshold higher, the high $Yb^{3+}$-doped $YAl_3(BO_3)_4$ and $GdAl_3(BO_3)_4$ crystals are potential candidates for efficient microchip laser media.

## 3.2. Microchip Dispensed with Processing

As mentioned above, thin microchip is essential to the sing longitudinal mode output and the efficient absorption of the LD pump source with low beam quality, but the processing of free-standing crystals is a great technological challenge, in particular for a thickness between 10 and 50μm. Therefore, the liquid phase epitaxy method has been used to obtain the thin microchip. A 48μm thin $KY(WO_4)_2$ layer doped with 20at.% $Yb^{3+}$ has been grown on a $KY(WO_4)_2$ substrate by the method with high optical quality and microchip laser operation of the Yb:$KY(WO_4)_2$/$KY(WO_4)_2$ epitaxial crystal has been achieved [79].

As another way to conquer the difficulty of microchip processing, the cleavage plates of laser crystals have attracted our attention. Lanthanum metaborate crystal $LaB_3O_6$ belongs to the monoclinic structure with space group $C2/c$ and is a biaxial crystal. The detailed crystallographic parameters of the crystal can be found in [80,81]. The X-ray structure studies of crystal show that the main structure unit of the $LaB_3O_6$ crystal is the boron-oxygen anion group $(B_6O_{12})^{6-}$ comprising two $BO_4$ tetrahedrons and four $BO_3$ triangles. These boron-oxygen groups are connected together through ionic bonds of $La^{3+}$ cations with nearest $O^{2-}$ ions and formed two-dimensional boron-oxygen layers parallel to the $(10\bar{1})$ planes [81]. When the ionic bonds between the neighboring layers break, the cleavage along boron-oxygen layers can take place [80,81]. Pure $LaB_3O_6$ crystals have been grown by the Czochralski method and plates with thickness of about 500μm can be prepared by cleaving. Such plates are suitable for manufacturing the scintillation detectors to record short-range radiation [80,81].

A 2.3at.% $Nd^{3+}$:$LaB_3O_6$ single crystal with high optical quality and the size up to $\phi 20\times 20mm^3$ was grown in our lab by the Czochraski method along a *b*-oriented seed. The grown $Nd^{3+}$:$LaB_3O_6$ crystal is shown in Fig. 5. The X-ray powder diffraction data have demonstrated that the crystal has the same crystallographic structure as the pure $LaB_3O_6$ crystal. The polarized absorption and fluorescence spectra were measured from the accurately oriented $Nd^{3+}$:$LaB_3O_6$ crystal at room temperature. Table 7 lists the structural and spectroscopic properties of the crystal [82]. It can be seen from the table that the spectroscopic parameters of $Nd^{3+}$:$LaB_3O_6$ crystal are comparable to those of $Nd^{3+}$:$YAl_3(BO_3)_4$ crystal. It indicates that the $Nd^{3+}$:$LaB_3O_6$ crystal is a promising material for solid-state laser.

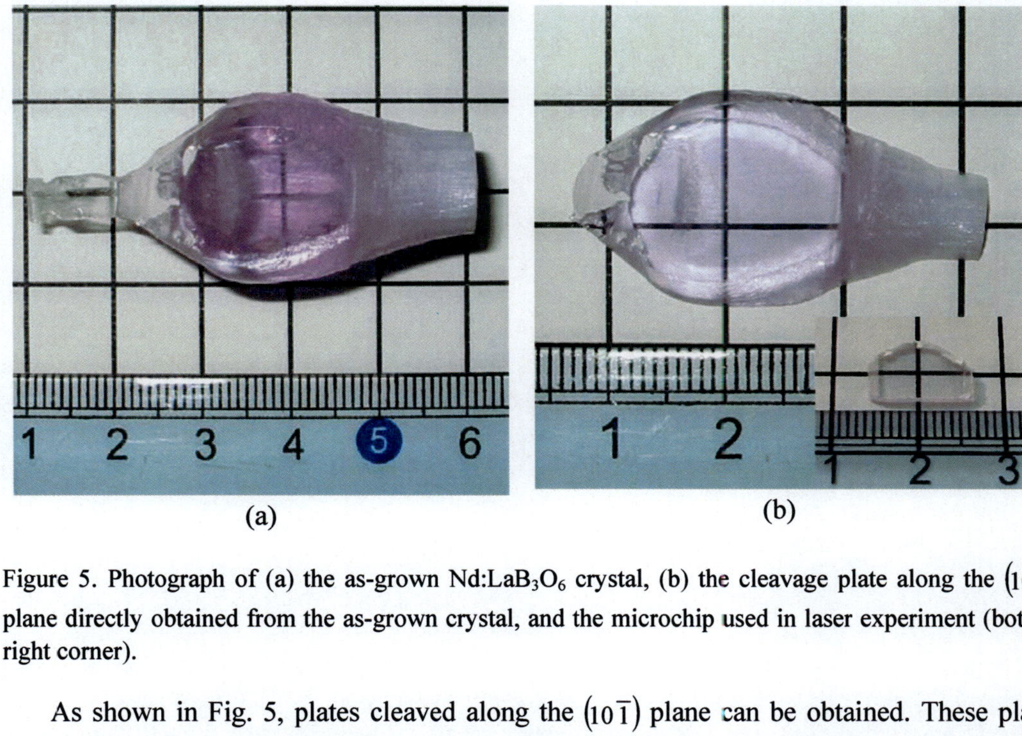


(a) (b)

Figure 5. Photograph of (a) the as-grown Nd:$LaB_3O_6$ crystal, (b) the cleavage plate along the $(10\bar{1})$ plane directly obtained from the as-grown crystal, and the microchip used in laser experiment (bottom right corner).

As shown in Fig. 5, plates cleaved along the $(10\bar{1})$ plane can be obtained. These plates have a high structural perfection without additional grinding, polishing, or etching. The thickness of the plate for microchip laser experiment, which is shown in the bottom right corner of Fig. 5b, is 1.0mm. Pumped by a Ti:sapphire laser at 871nm, 1060nm continuous-wave laser emission with slope efficiency of 23% has been achieved in the unprocessed microchip [83]. The result cannot compare with those of other microchip lasers yet, such as widely investigated $Nd^{3+}$:YAG and $Nd^{3+}$:$YVO_4$ [1,5,18]. It can be attributed to the following reasons.

**Table 7. Structural and spectroscopic parameters of Nd:$LaB_3O_6$ crystal**

| | |
|---|---|
| Crystal system | monoclinic biaxial |
| Space group | *C2/c* (or *I2/a*) |
| Growth method | Czochralski |
| $Nd^{3+}$ concentration ($10^{20}cm^{-3}$) | 2.18 |
| Peak absorption wavelength $\lambda_a$ | 799nm |
| FWHM at $\lambda_a$ | 3nm |
| $\sigma_a$ ($10^{-20}cm^2$) | 5.03 (*E//X*)<br>4.51 (*E//Y*)<br>5.05 (*E//Z*) |
| Peak fluorescence wavelength $\lambda_e$ | 1062.8nm (*E//X*)<br>1054.7nm (*E//Y*)<br>1062.8nm (*E//Z*) |
| FWHM at $\lambda_e$ | 4nm |
| $\sigma_e$ ($10^{-20}cm^2$) | 10.17 (*E//X*)<br>8.29 (*E//Y*)<br>7.22 (*E//Z*) |
| Radiative lifetime of $^4F_{3/2}$ | 302μs |
| Fluorescence lifetime of $^4F_{3/2}$ | 53μs |
| Fluorescence quantum efficiency of $^4F_{3/2}$ | 0.18 |

Firstly, the concentration of $Nd^{3+}$ ions used in this experiment is not optimal for the $Nd^{3+}$:$LaB_3O_6$ microchip with 1.0mm thickness because only 50% incident pump power is absorbed. With the increase of the $Nd^{3+}$ concentration in the crystal, it can be anticipated that a thinner microchip can be used in the laser experiment and the better laser performance can be realized. Secondly, more cooling techniques must be employed to reduce the pump-induced thermal load, which caused the degradation of the laser performance and the instability of the output power, especially for the microchip lasers [19,43,45]. Thirdly, the configuration of the laser resonator must be improved.

## 4. CONCLUSION

In this chapter, taking into account the crystal growth and the fluorescence quenching, several commercial $Nd^{3+}$ and $Yb^{3+}$-doped laser crystals, especially, the Nd:YAG, Nd:$YVO_4$, Yb:YAG, and Yb:$KGd(WO_4)_2$, have been analyzed as microchip laser media. The chemical structure reasons for weak concentration dependent fluorescence quenching in the borate laser crystals have been discussed. The crystals grown in our lab suitable for microchip laser, including stoichiometric laser crystal $YbAl_3(BO_3)_4$, and crystals of $YAl_3(BO_3)_4$ and $GdAl_3(BO_3)_4$ doped with a series of $Yb^{3+}$ concentrations, have been introduced.

Two methods for conquering the difficulty of microchip processing, i.e. the liquid phase epitaxy and the cleavage plates have been introduced. Although there are still a lot of works required to be done before the commercial use of the unprocessed cleavage plates as microchip laser gain media, the initial work proposes a new and easier method for obtaining the microchip laser gain medium and this method can also be applied to other cleavage laser crystals with better laser properties.

Besides the fundamental and the intra-cavity Q-switched and/or frequency conversion microchip lasers, the self-frequency-doubling and the Raman self-frequency conversion microchip lasers have also been reported. The former is based on the laser crystals with nonlinear optical property and a visible laser can emit from a single so-called nonlinear laser microchip pumped by LD [68,84]. The latter is based on the laser crystals with large Raman gain coefficient and the wavelength of output laser can be shifted by the stimulated Raman scattering of the microchip itself [85]. Furthermore, the self-tunable microchip laser is also possible [35]. Through using the multifunctional crystals, the absolutely monolithic microchip devices with visible and other practical wavelength laser emission can be realized.

## ACKNOWLEDGMENTS

We thank all our coworker and our experimental colleagues J. S. Liao, X. Q. Lin, Y. J. Chen, Z. Y. Huang, and Q. G. Tan for fruitful cooperation. Financial supported by the National Natural Science Foundation of China (Grant No. 50590405 and 50372068) and the Major Programs of Science and Technology Foundation of Fujian Province (Grant No. 2002H004 and 2004HZ01-1) is gratefully acknowledged.

## REFERENCES

[1] Zayhowski, J. J.; Mooradian, A. *Opt. Lett.* 1989, 14, 24-26.
[2] Zayhowski, J. *J. Opt. Mater.* 1999, 11, 255-267.
[3] Molva, E. *Opt. Mater.* 1999, 11, 289-299.
[4] Sinclair, B. D. *Opt. Mater.* 1999, 11, 217-233.
[5] Sasaki, T.; Kojima, T.; Yokotani, A.; Oguri, O.; Nakai S. *Opt. Lett.* 1991, 16, 1665-1667.
[6] Spuhler, G. J.; Paschotta, R.; Kulloerg, M. P.; Graf, M.; Moser, M.; Mix, E.; Huber, G.; Harder, C.; Keller, U. *Appl. Phys. B* 2001, 72, 285-287.
[7] Koechner, W. *Solid-State Laser Engineering*; Spring-Verlag: Berlin, 1999.
[8] Beier, B.; Meyn, J.-P.; Knappe, R.; Boller, K.-J.; Huber, G.; Wallenstein, R. *Appl. Phys. B*, 1994, 58, 381-388.
[9] Wyss, Chr. P.; Luthy, W.; Weber, H. P.; Vlasov, V. I.; Zavartsev, Yu. D.; Studenikin, P. A.; Zagumennyi, A. I.; Shcherbakov, I. A. *Appl. Phys. B*, 1999, 68, 659-661.
[10] Spuhler, G. J.; Paschotta, R.; Fluck, R.; Braun, B.; Moser, M.; Zhang, G.; Gini, E.; Keller, U. *J. Opt. Soc. Am. B*, 1999, 16, 376-388.
[11] Dong, J.; Deng, P.; Gan, F.; Urata, Y.; Hua, R.; Waba, S.; Tashiro, H. *Opt. Commun.* 2001, 197, 413-418.
[12] Taira, T.; Mikai, A.; Nozawa, Y.; Kobayashi, T. *Opt. Lett.* 1991, 16, 1955-1957.
[13] Hur, M. G.; Yang, W. S.; Suh, S. J.; Ivanov, M. A.; Kochurikhin, V. V.; Yoon, D. H. *J. Cryst. Growth* 2002, 237-239, 745-748.
[14] Lupei, V.; Pavel, N.; Taira, T. *Opt. Lett.* 2001, 26, 1678-1680.

[15] Kintz, G. J.; Baer, T. *IEEE J. Quantum Electron.* 1990, 26, 1457-1459.
[16] Helmfrid, S.; Tastumo, K. *J. Opt. Soc. Am. B* 1994, 11, 436-445.
[17] Lavi, R.; Jackel, S.; Tal, A.; Lebiush, E.; Tzuk, Y.; Goldring, S. *Opt. Commun.* 2001, 195, 427-430.
[18] Sato, Y.; Taira, T.; Pavel, N.; Lupei, V. *Appl. Phys. Lett.* 2003, 82, 844-846.
[19] Luo, Z.; Huang, Y.; Montes, M.; Jaque, D. *Appl. Phys. Lett.*2004, 85, 715-717.
[20] Lupei, V.; Pavel, N.; Taira, T. *Appl. Phys. Lett.* 2002, 81, 2677-2679.
[21] Yaney P.; Deshazer, L. *J. Opt. Soc. Am.*1976, 66, 1405-1414.
[22] Sardar, D. K.; Castano, F.; French, J. A.; Gruber, J. B.; Alekel, T.; Keszler, D. A.; Clark, B. L. *J. Appl. Phys.* 2001, 90, 4997-5001.
[23] Lupei, V.; Lupei, A.; Pavel, N.; Taira, T.; Shoji, I.; Ikesue, A. *Appl. Phys. Lett.* 2001, 79, 590-592.
[24] Huang, Z.; Huang, Y.; Chen, Y.; Luo, Z. *J. Opt. Soc. Am. B* 2005, in press.
[25] Lupei, V.; Pavel, N.; Taira, T. *Opt. Commun.* 2002, 201, 431-435.
[26] Chinn, S. R.; Hong, H. Y.-P.; Pierce, J. W. *Laser Focus* May 1976, 12, 64-69.
[27] Luo, Z.-D.; Jiang, A.-D.; Huang, Y.-C.; Qiu, M.-W. *Sci. in China A* 1991, 34, 762-768.
[28] Asakawa, Y.; Kawai, R.; Ohki, K.; Otsuka, K. Jpn. *J. Appl. Phys.* 1999, 38, L515-L517.
[29] Jaque, D.; Enguita, O.; Garcia Sole, J.; Jiang, A. D.; Luo, Z. D. *Appl. Phys. Lett.* 2000, 76, 2176-2178.
[30] Luo, Z.; Huang, Y.; Montes, M.; Jaque, D. *Appl. Phys. Lett.* 2004, 85, 715-717.
[31] Kaminskii, A. A.; Sarkisov, S. E.; Bohn, J.; Reiche, P.; Schultze, D.; Uecker, R. *Phys. Stat. Sol. (a)* 1977, 43, 71-79.
[32] Chinn, S. R.; Hong, H. Y.-P. *Opt. Commun.* 1975, 15, 345-350.
[33] Miyazawa, S.; Kubodera, K. *J. Appl. Phys.* 1978, 49, 6197-6198.
[34] Choy, M. M.; Zwicker, W. K.; Chinn, S. R. *Appl. Phys. Lett.* 1979, 34, 387-388.
[35] Fernandez, J.; Iparraguirre, I.; Aramburu, I.; Illarramendi, A.; Azkargorta, J.; Voda, M.; Balda, R. *Opt. Lett.* 2003, 28, 1341-1343.
[36] Auzel, F. *Mater. Res. Bull.* 1979, 14, 223-231.
[37] Gueugnon, C.; Budin, J. P. *IEEE J. Quantum Electron.* 1980, 16, 94-98.
[38] Auzel, F. *Opt. Mater.* 2002, 19, 89-94.
[39] Jaque, D.; Enguita, O.; Caldino, U.; Ramirez, M. O.; Garcia Sole, J.; Zaldo, C.; Munoz-Santiuste J. E.; Jiang, A. D.; Luo, Z. D. *J. Appl. Phys.* 2001, 90, 561-569.
[40] Jaque, D.; Enguita, O.; Caldino, G. U.; Ramírez, M. O.; Garcia Sole, J.; Zaldo, C.; Munoz-Santiuste, J. E.; Jiang, A. D.; Luo, Z. D. *J. Appl. Phys.* 2001, 90, 561-569.
[41] Krupke, W. F. *IEEE J. Selected Topic Quantum Electron.* 2000, 6, 1287-1296.
[42] Brenier, A.; Boulon, G. *J. Alloy. Compound.* 2001, 323-324, 210-213.
[43] Patel, F. D.; Honea, E. C.; Speth, J.; Payne, S. A.; Hutcheson, R.; Equall, R. *IEEE J. Quantum Electron.* 2001, 37, 135-144.
[44] Demidovich, A. A.; Kuzmin, A. N.; Ryabtsev, G. I.; Danailov, M. B.; Strek, W.; Titov, A. N. *J. Alloy. Compd.* 2000, 300-301, 238-241.
[45] Klopp, P.; Griebner, U.; Petrov, V.; Mateos, X.; Bursukova, M. A.; Pujol, M. C.; Sole, R.; Gavalda, J.; Aguilo, M.; Guell, F.; Massons, J.; Kirilov, T.; Diaz, F. *Appl. Phys. B* 2002, 74, 185-189.
[46] Liao, J.; Lin, Y.; Chen, Y.; Luo, Z.; Huang, Y. *J. Cryst. Growth* 2004, 267, 134-139.
[47] Liao, J.; Lin, Y.; Chen, Y.; Luo, Z.; Huang, Y. *J. Alloy. Compd.* 2005, 397, 211-215.

[48] Gaume, R.; Haumesser, P. H.; Viana, B.; Vivien, D.; Ferrand, B.; Aka, G. *Opt. Mater.* 2002, 19, 81-88.

[49] Druon, F.; Chenais, S.; Raybaut, P.; Balembois, F.; Georges, P.; Gaume, R.; Aka, G.; Viana, B.; Mohr, S.; Kopf, D. *Opt. Lett.* 2002, 27, 197-199.

[50] Chenais, S.; Druon, F.; Balembois, F.; Georges, P.; Gaume, R.; Haumesser, P. H.; Viana, B.; Aka, G. P.; Vivien, D. *J. Opt. Soc. Am. B*, 2002, 19, 1083-1090.

[51] Bayramian, A. J.; Marshall, C. D.; Schaffers K. I.; Payne, S. A. *IEEE J. Quantum Electron.* 1999, 35, 665-674.

[52] Lagatsky, A. A.; Abdolvand, A.; Kuleshov, N. V. *Opt. Lett.* 2000, 25, 616-618.

[53] Druon, F.; Chenais, S.; Balembois, F.; Georges, P.; Brun, A.; Courjaud, A.; Honninger, C.; Salin, F.; Zavelani-Rossi, M.; Auge, F.; Chambaret, J. P.; Aron, A. Mougel, F.; Aka, G.; Vivien, D. *Opt. Mater.* 2002, 19, 73-80.

[54] Yang, P.; Deng, P.; Yin, Z. *J. Lumin.* 2002, 97, 51-54.

[55] Qiu, H.; Yang, P.; Dong, J.; Deng, P.; Xu, J.; Chen, W. *Mater. Lett.* 2002, 55, 1-7.

[56] Auzel, F. *J. Lumin.* 2002, 100, 125-130.

[57] Mukhopadhyay, P. K.; George, J.; Ranganathan, K.; Sharma, S. K.; Nathan, T. P. S. *Opt. Laser Tech.* 2002, 34, 253-258.

[58] Pujol, M. C.; Bursukova, M. A.; Guell, F.; Mateos, X.; Sole, R.; Gavalda, J.; Aguilo, M.; Massons, J.; Diaz, F. *Phys. Rev. B* 2002, 65, 165121.

[59] Auzel, F.; Bonfigli, F.; Gagliari, S.; Baldacchini, G. *J. Lumin.* 2001, 94-95, 293-297.

[60] Sumida, D. S.; Fan, T. Y. *Opt. Lett.* 1994, 19, 1343-1345.

[61] Takebe, H.; Murata, T.; Morinage, K. *J. Am. Ceram. Soc.* 1996, 79, 681-687.

[62] Zou, X.; Toratani, H. *Phys. Rev. B* 1995, 52,, 15889-15897.

[63] Huang, Z.; Huang, Y.; Huang, M.; Luo, Z. *J. Opt. Soc. Am. B* 2003, 20, 2061-2067.

[64] Huang, Z.; Gong, X.; Huang, Y.; Luo, Z. *Opt. Commun.* 2004, 237, 389-397.

[65] Luo, Z.-D. Chem. *Phys. Lett.* 1983, 94, 498-504.

[66] Luo, Z.; Jiang, A.; Huang, Y.; Qiu, M. Chinese *Phys. Lett.* 1989, 6, 440-443.

[67] Tu, C.; Qiu, M.; Huang, Y.; Chen, X.; Jiang, A.; Luo, Z. *J. Cryst. Growth* 208, 2000, 487-492.

[68] Burns, P. A.; Dawes, J. M.; Dekker, P.; Piper, J. A.; Li, J.; Wang, J. *Opt. Commun.* 2002, 207, 315-320.

[69] Wang, P.; Dawes, J. M.; Dekker, P.; Piper, J. A. *Opt. Commun.* 2000, 174, 467-470.

[70] Leonyuk, N. I.; Leonyuk, L. I. *Prog. Cryst. Growth Character* 1995, 31, 179-278.

[71] Jiang, H.; Li, J.; Wang, J.; Hu, X.; Liu, H.; Teng, B.; Zhang, C.; Dekker, P.; Wang, P. *J. Cryst. Growth* 2001, 233, 248-252.

[72] Leonyuk, N. I.; Koporulina, E. V.; Barilo, S. N.; Kurnevich, L. A.; Bychkov G. L. *J. Cryst. Growth* 1998, 191, 135-142.

[73] Koporulina, E. V.; Leonyuk, N. I.; Pilipenko, O. V.; Mokhov, A. V.; Bocelli, G.; Righi, L. *Mater. Lett.* 2001, 47, 145-149.

[74] Xu, Y.; Gong, X.; Chen, Y.; Huang, M.; Luo, Z.; Huang, Y. *J. Cryst. Growth* 2003, 252, 241-245.

[75] Xu, Y.-Y.; Chen, Y.-J.; Luo, Z.-D.; Chen, J.-T.; Huang, Y.-D. Chinese *J. Struct. Chem.* 2002, 21, 402-404.

[76] Mills, A. D. *Inorg. Chem.* 1962, 1, 960-961.

[77] Wang, P.; Dawes, J. M.; Dekker, P.; Knowles, D. S.; Piper, J. A. *J. Opt. Soc. Am. B* 1999, 16, 63-69.

[78] Auzel, F. in: *Spectroscopy of solid-state laser-Type materials*; Di Bartolo, B.; Ed.; Plenum Press: New York, 1987, pp293-341.

[79] Aznar, A.; Sole, R.; Aguilo, M.; Diaz, F.; Griebner, U.; Grunwald, R.; Petrov, V. *Appl. Phys. Lett.* 2004, 85, 4313-4315.

[80] Shekhovtsov, A. N.; Tolmachev, A. V.; Dubovik, M. F.; Korshikova, T. I.; Baumer, V. N.; Grinyov, B. V.; Dolzhenkova, E. F. *Nucl. Instrum. Methods Phys. Res. A* 2001, 456, 280-283.

[81] Dolzhenkova, E. F.; Shekhovtsov, A. N.; Tolmachev, A. V.; Dubovik, M. F.; Grinyov, B. V.; Tarasov, V. A.; Baumer, V. N.; Zelenskaya, O. V. *J. Cryst. Growth* 2001, 233, 473-476.

[82] Chen, Y.; Lin, X.; Luo, Z.; Huang, Y. Chem. *Phys. Lett.* 2004, 397, 282-287.

[83] Chen, Y. J.; Huang, Y. D.; Lin, X. Q.; Tan, Q. G.; Luo, Z. D.; Lin, Y. F. *Appl. Phys. Lett.* 2005, 86, 021115.

[84] Aka, G.; Reino, E.; Loiseau, P.; Vivien, D.; Ferrand, B.; Fulbert, L.; Pelenc, D.; Lucas-Leclin, G.; Georges, P. *Opt. Mater.* 2004, 26, 431-436.

[85] Grabtchikov, A. S.; Kuzmin, A. N.; Lisinetskii, V. A.; Orlovich, V. A.; Demidovich, A. A.; Danailov, M. B.; Eichler, H. J.; Bednarkiewicz, A.; Strek, W.; Titov, A. N. *Appl. Phys. B* 2002, 75, 795-797.

In: Trends in Crystal Growth Research
Editor: George V. Karas, pp. 45-69
ISBN 1-59454-541-3


*Chapter 3*

# FEASIBILITY OF GROWING AND OPTICAL PROPERTIES OF FERROELECTRIC (1-X)PB($MG_{1/3}NB_{2/3}$)$O_3$-XPBTIO$_3$ SINGLE CRYSTALS

***Xinming Wan, Haosu Luo, Xuehua Zhang and Xiangyong Zhao***
The State Key Laboratory of High Performance Ceramics and Superfine Microstructure, Shanghai Institute of Ceramics, Chinese Academy of Sciences, 215 Chengbei Road Jiading, Shanghai 201800, China

## ABSTRACT

Relaxor ferroelectric $(1\text{-}x)Pb(Mg_{1/3}Nb_{2/3})O_3\text{-}xPbTiO_3$ (PMN-PT) single crystals have come into prominence due to their extrahigh electromechanical coupling factor ($k_{33}$ ~ 94%), piezoelectric coefficient ($d_{33}$ ~ 2500 pC/N), dielectric constant ($\varepsilon_r$ ~ 5000 ~ 5500) and field-induced strain response ($S$ ~ 1.7%). Many techniques have been used to grow PMN-PT single crystals, such as flux growth (high temperature solution method), melt growth (high-pressure Czochralski and Bridgman growth) and hydrothermal growth. Though most $PbTiO_3$ (PT)-based single crystals can be grown from a PbO flux, it is difficult to grow large-size and good-quality PMN-PT single crystals in a production-viable and cost-effective way. On the basis of the study of the phase stability and phase structure, the feasible method is investigated and it is verified by our researches that PMN-PT single crystals can be grown directly from melt for their stability of perovskite phase at high temperature, so a modified Bridgman technique is used to grow PMN-PT single crystals. By the B-site precursor synthesis method and using seed crystals, the crystal boules with 50 mm in diameter and about 80 mm in length have been obtained in our group. A mass of studies had been carried out on their unusually large piezoelectric response, while in this chapter we will report their up-to-date optical properties. From the results of the refractive indices, the optical transmittance, the extinction ratio and the linear electro-optic (EO) properties, we have found that PMN-PT single crystals are very promising candidates for EO materials.

# I. INTRODUCTION

For the past several decades, $Pb(Zr_{1-x}Ti_x)O_3$ (PZT)-based ceramics have been widely used in actuator, transducer, et al. In recent years, $PbTiO_3$ (PT)-based single crystals in perovskite structure of complex $Pb(B'B'')O_3$-$PbTiO_3$ have been found to exhibit electromechanical properties superior to those of the conventional PZT-based ceramics and have attracted much attention. In 1997, Park and Shrout [1] found that $(1-x)Pb(Mg_{1/3}Nb_{2/3})O_3$-$xPbTiO_3$ (PMN-PT) and $(1-x)Pb(Zn_{1/3}Nb_{2/3})O_3$-$xPbTiO_3$ (PZN-PT) single crystals exhibited very high piezoelectric coefficients ($d_{33}$ ~ 2500 pC/N), dielectric constant ($\varepsilon_r$ ~ 5000 ~ 5500) and field-induced strain response ($S$ ~ 1.7%) with little hysteresis along the [001] direction. As a result, these materials are a group of promising materials for various applications, such as ultrasound transducers and high strain actuators [1-3]. The piezoelectric properties of these materials peak near the morphotropic phase boundary (MPB) where the solid solution changes structure from rhombohedral to tetragonal, near 33% PT ($x$ = 0.33) in PMN-PT and near 8% PT ($x$ = 0.08) in PZN-PT. Although PZN-PT is found to have higher piezoelectric values [1], PMN-PT has a more stable perovskite phase and exhibits less of a tendency to form the parasitic pyrochlore phase, making it easier to produce [4,5].

Single crystal $Pb(Mg_{1/3}Nb_{2/3})O_3$ was first reported by Smolenskii et al. in 1960 [6]. Subsequently, the single crystal growth and electrical properties of PMN and PMN-PT have been investigated in depth [1-14]. Many researchers have managed to use the flux technique to grow PMN-PT single crystals [1,5,7,8]. However, because of the difficulties of controlling spontaneous nucleation, the growth results are difficult to reproduce, with many grown crystals containing secondary pyrochlore phases and inclusions of the PbO flux, which seriously decreases the dielectric and piezoelectric properties of the ferroelectric system. Moreover, this method is not conducive to the growth of large-size and good-quality single crystals which are required for commercial devices. Besides size and orientation, loss of material due to the high volatility of PbO containing melts is also a major concern when choosing an appropriate bulk growth technique for this system. Therefore, in order to improve the growth quality and stability of PMN-PT single crystals, the control of nucleation and the stabilization of the perovskite phase, by both the thermal and kinetic methods, are very important factors.

Luo et al. have published a lot of papers on the modified Bridgman growth of PMN-PT single crystals [9-14]. They described problems with the crystal quality including the formation of dendrites [10,13], inclusions [14], voids [14] and compositional variations [10,12,14]. They also confirmed that the Pt crucibles were prone to leaking [12]. Little information was provided concerning specific growth parameters related to the crystal quality [9,10,12].

In this chapter, the feasible method will be investigated on the basis of the study of the phase stability and phase structure, and it has been verified by our researches that PMN-PT single crystals can be grown directly from melt for their stability of perovskite phase at high temperature, so a modified Bridgman technique with the B-site precursor synthesis method is used to grow PMN-PT single crystals. The morphology and growth mechanism of PMN-PT single crystals are also discussed.

Most oxygen-octahedra ferroelectrics, which exhibit excellent electromechanical properties, also have outstanding optical properties [15,16]. Some applications, such as photorefractive, second-harmonic generation (SHG), electro-optic (EO), and elasto-optical devices, are based on the large optical property coefficients of the material [17,18]. PMN-PT single crystals are the materials very well known for their high electromechanical coupling factor and numerous studies had been carried out on their dielectric constant, piezoelectric coefficient and field induced strain response [1-4,9,10]. However, in spite of this extensive utilization, the values of the optical properties of PMN-PT single crystals are not perfectly known. The knowledge of the detailed optical parameters, such as the refractive indices or optical transmission properties, is most desirable, not only for making possible practical usage in optic devices but also to answer very basic questions on the energy level of the crystals and the physical nature of this class of relaxor ferroelectrics.

In this chapter, a systematical study of the optical properties for PMN-PT single crystals will be carried out, instead of electrical properties which had been reported widely. From the up-to-date results of the refractive indices, the optical transmittance, the extinction ratio and the linear EO properties, we have found that PMN-PT single crystals are very promising candidates for EO materials.

## II. Crystal Growth

### A. Feasibility of the Growth of Pt-Based Single Crystals

It is very important to study the feasibility of the crystal growth if one hopes to establish a technique to obtain single crystals with large-size and high-quality. For solid solution PT-based single crystals, it is very necessary to study how to control the volatility of oxides and phase stability of perovskite at high temperatures.

For the growth of PMN-PT single crystals, there are at least four kinds of oxides used as raw materials. Among them, PbO melts only at 886 °C and it is easily volatile during the crystal growth because PT-based single crystals usually grow at the temperature higher than 1200 °C. For example, the melting point is about 1320 °C, 1296 °C, 1288 °C, 1284 °C, 1255 °C for PMN, PMN-0.30PT, PMN-0.35PT, PMN-0.40PT, PT, respectively [9,14], and about 1250 °C for PZN-0.09PT [19]. The volatility of oxides may result in not only nonstoichiometry and increased defects in single crystals, but also environmental contamination, so a feasible method for the growth of PT-based single crystals should seal raw materials in an isolated system to prevent the evaporation of PbO during the crystal growth and ensure the certain composition for crystal growth. For conventional Czochralshi method, it is difficult to prevent the evaporation of PbO during crystal growth, so this method is difficult for the growth of PT-based single crystals.

Some PT-based single crystals are thermodynamic unstable at high temperature. For example, we were failed to grow large-size PZN-PT single crystals directly from melt because the pyrochlore phases easily appear during the crystal growth. However, most PT-based single crystals can be grown from high temperature solution by using PbO flux, which will increase the stability of PT-based single crystals and restrain the appearance of pyrochlore phases. Yamashita et al. have confirmed that flux Bridgman method is suitable for growing PZN-PT single crystals [20,21]. Fang and Luo have obtained PZN-PT single crystals by a modified Bridgman technique with PbO flux using an allomeric seed crystal [22,23]. But a process of removing PT-based single crystals out from PbO flux is necessary after finishing the crystal growth process.

Pyrochlore phases easily appear during the growth of PT-based single crystals because PMN-PT single crystals are thermodynamic unstable at high temperature and they easily lose MgO and PbO during the crystal growth. For example, the pyrochlores $Pb_6MgNb_6O_{22}$ ($P_6MN_3$), $Pb_3Nb_2O_8$ ($P_3N$), $Pb_2Nb_2O_7$ ($P_2N$), $Pb_{1.5}Nb_2O_{6.5}$ ($P_3N_2$) and $Pb_5Nb_4O_{15}$ ($P_5N_2$) often appear on the boule surface of as-growth PMN-PT single crystals, where the PbO and MgO easily disappear during the crystal growth due to the evaporation of PbO and instability of PMN-PT at high temperature. However, it has been verified that PMN-PT single crystals could be grown directly from melt if a sealed platinum crucible is used to prevent the evaporation of PbO and melting temperature is much higher than its incongruent temperature to facilitate the complete reaction of oxides during the crystal growth [9,11,12,24].

In general, crystals grow very slowly from a flux because solvent stunts the process of mass transportation during the crystal growth. For example, we grow the high efficient photorefractive $BaTiO_3$ single crystals by the top-seeded solution growth (TSSG) method with the rate of about 0.1 mm/h, which is much slower than that of the growth method directly from melts.

To compare the crystal growth directly from melt, it is clear that the flux method for the growth of large-size PT-based single crystals is not a production-viable way due to its lower growth rate and higher cost of materials and energy. Therefore, the growth method directly from melt is suitable for mass-producing large-size PT-based single crystals for practical applications. The flux method is used only for those crystals cannot be grown directly from melt. For example, PZN-PT single crystals could only be grown from flux, generally from PbO flux.

Many PT-based single crystals have been successfully grown in our laboratory on the basis of the researches on thermodynamic stability and phase relationship of PT-based system, such as PMN-PT [9-14], PZN-PT [19,22,23,25], $Pb(Sb_{1/2}Nb_{1/2})O_3$-$Pb(Mg_{1/3}Nb_{2/3})O_3$-$PbTiO_3$ (PSMNT) [26], $Pb(Fe_{1/2}Nb_{1/2})O_3$-$PbTiO_3$ (PFN-PT) [27] and $Pb(In_{1/2}Nb_{1/2})O_3$-$PbTiO_3$ (PIN-PT) [28,29]. Among them, PMN-PT, PSMNT, PFN-PT and PIN-PT single crystals can be grown directly from their high-temperature melt. However, PZN-PT single crystals can only be grown from their high-temperature solution because of the instability of the perovskite phase at high temperature.

A principle has been adopted to grow large-size and high-quality PT-based single crystals in our laboratory. That is to say, the melt growth method is chosen firstly to avoid the flux growth except for these PT-based single crystals which could not be manufactured directly from melt. Our results have verified that the Bridgman method is a feasible technique to grow PMN-PT, PSMNT, PFN-PT and PIN-PT single crystals in a cost-effective and production-viable way. After optimizing the growth process, we have successfully obtained large-size and high-quality PMN-PT single crystals for commercial applications.

## B. Material Synthesis

A modified Bridgman technique is used to grow ferroelectric PMN-PT single crystals on the basis of investigation of the phase relation and phase stability of PMN-PT binary system [9-11,30]. The raw powders of PbO, MgO, $Nb_2O_5$ and $TiO_2$ with purity more than 99.99% are used as the starting materials. High-purity starting materials should be used since the mixed dissociative metal may react with Pt as well and form Pt-metal alloys with lower melting point. The experiments have demonstrated that these measures are effective. The leak of Pt crucibles can be suppressed to a great extent and the starting materials almost have no gravity loss.

Some defects (e.g. dendrites, inclusions and voids) have been found in PMN-PT single crystals grown by directly melting the component oxides [9-14] and they may be caused by melt inhomogeneity. A homogeneous PMN-PT melt may be difficult to form from the component oxides because of the large differences in melting temperature and density. For example, MgO is not only more refractory than PbO ($T_{m\text{-}MgO}$ = 2852 °C, $T_{m\text{-}PbO}$ = 886 °C [31]), but also much lighter ($\rho_{MgO}$ = 3.58 g/cm$^3$, $\rho_{PbO}$ = 9.53 g/cm$^3$ [31]). Molten PbO easily decomposes into lead at high temperature and forms lead-platinum alloy to cause leakage of platinum crucible, resulting in deviation of batch composition. In order to avoid melt inhomogeneity problems, the component oxides of raw materials are precalcined by the B-site precursor synthesis method, a modified version of the columbite precursor method [32]. The method is also favorable for the stability of pure perovskite phase and restraining the pyrochlore phase during crystal growth. In the method, raw powders of MgO and $Nb_2O_5$ are presynthesized in the reaction:

$$MgO + Nb_2O_5 \rightarrow MgNb_2O_6 \quad (1)$$

Then the resulting mixture is ground into powder by $Al_2O_3$ mortar. With the addition of specified stoichiometric PbO and $TiO_2$, the resultant is mixed in ball-milling equipment with $ZrO_2$ balls and distilled water for 24 h. After parching and milling, the resulting mixture is sealed in cylindrical platinum crucibles to prevent the evaporation of PbO and control the uniformity of composition during crystal growth. The growth procedure will be delineated in detail in the next paragraph.

## C. Growth Procedure

PMN-PT single crystals are grown by a modified Bridgman technique. Figure 1 shows the schematic diagram of the modified Bridgman furnace. The preparation of platinum crucibles is very important in the Bridgman growth of PMN-PT single crystals since PbO containing melts can attack Pt at elevated temperatures. Leaks of any amount can be a problem and large leaks can ruin an experiment completely. The best results are found with the crucibles made from Pt foil which have three layers and about 0.15 mm thick for each layer. After filled with raw materials, Pt crucibles should be sealed and then placed inside alumina powder, as shown in Fig. 1b and c.

The orientation growth technique is adopted to effectively enlarge the size of PMN-PT single crystals. The anisotropy in growth rates appears evidently in the growth of PMN-PT single crystals, which is favorable for the washing out of polycrystals. It was observed that {111} orientation predominated in the cuts normal to the growth direction or the longitudinal one of cylinder boules while no crystal seed was used. This crystal habit is used to the orientation growth of PMN-PT single crystals. So, the seed crystals are usually used during crystal growth, along <111> direction, as shown in Fig. 1e, and they have a key effect on restraining spontaneous nucleation and parasitic growth.

A resistance furnace with $MoSi_2$ heating elements, as shown in Fig. 1a, is used for growth. During the crystal growth, the highest solution temperature is usually more than 100 °C above its melting point (depending on their composition) and the temperature gradient is about 30 - 50 °C/cm normal to the solid-liquid interface. The temperature in the Bridgman furnace is regulated by a proportional integral differential (PID) controller and the precision of the furnace temperature is controlled within ± 0.5 °C. The growth temperature profile used is rapidly heated to the temperature more than 100 °C above its melting point, and then after soaking for about 10 h, the crucible is pulled down automatically at a rate of 0.1-1.0 mm/h by a drop-down mechanism, as shown in Fig. 1d. The descending rate of Pt crucibles is engineered to make the shape of solid-liquid interfaces be slightly convex and their positions keep stable where the temperature gradient is the largest on the temperature curves of the furnaces. The crucibles are air cooled at the bottom and unidirectional solidification is accomplished through either furnace or crucible translation. At the end process of growth, the furnace temperature is decreased slowly to prevent cracking, usually at a rate of 25 °C/h. Particular care should be taken in the range of 200 °C to room temperature (cooling rate of 5 °C/h) where the material undergoes a phase transition from cubic to tetragonal or rhombohedral (depending on composition). As last, the Pt crucibles are bathed with acid and weighed to evaluate the weight loss of contents during the crystal growth.

During the crystal growth, a vapor filled region forms above the melt because the charges, composed of packed powder, shrink on melting. Pressure builds up in this region at growth temperatures, and the upper region ~ 2 - 3 cm of the crucibles typically swells by several millimeters. The swelling usually does not affect the growth run. However, small amounts of orange or red material are often deposited at the top of the crucibles. The deposition is the evaporated PbO and it induces the segregation during the crystal growth. Because of segregation, PT content increases from the bottom to the top, i.e. increases with the distance from the seed crystal. The content of PT can be measured from x-ray fluorescence analysis. $d_{(hkl)}$ values from x-ray diffraction (XRD) pattern also showed that the variation of composition during the growth of PMN-PT single crystals with the initial composition near the MPB. For example, in the growth of PMN-0.33PT single crystal, XRD patterns are different between the bottom and top cut from the same boule [10]. The peaks of (110) split into two peaks at the top part of the boule while at the bottom part, there is only one peak. It shows that the phase transition from rhombohedral to tetragonal is happened, indicating the increasing of PT content from bottom to top of the same boule. The segregation can also be confirmed by measuring Curie temperature $T_c$ of the samples from different parts of one boule [33]. The result shows that the lower $T_c$ value is obtained at the bottom part and higher $T_c$ at the top part of the same boule.

The segregation during crystal growth brings the compositional variation in a same crystal boule. We can not eliminate the segregation effect completely due to the lower melting temperature of PbO (886 °C) compared with the growth temperature, usually above 1300 °C [10,24], but the segregation is not severe since the raw materials are sealed in an isolated Pt crucible. Now, we can control the effective segregation coefficient $k_e$ within $0.95 < k_e < 1$ [10,24]. The case of $k_e < 1$ is in agreement with the slope of liquid phase line of PMN-PT binary system [10], and the value of $k_e$ closes to 1 means that the PMN-PT single crystals can be regarded as a near congruently melting compound, which can be crystallized from melt directly by the modified Bridgman method.

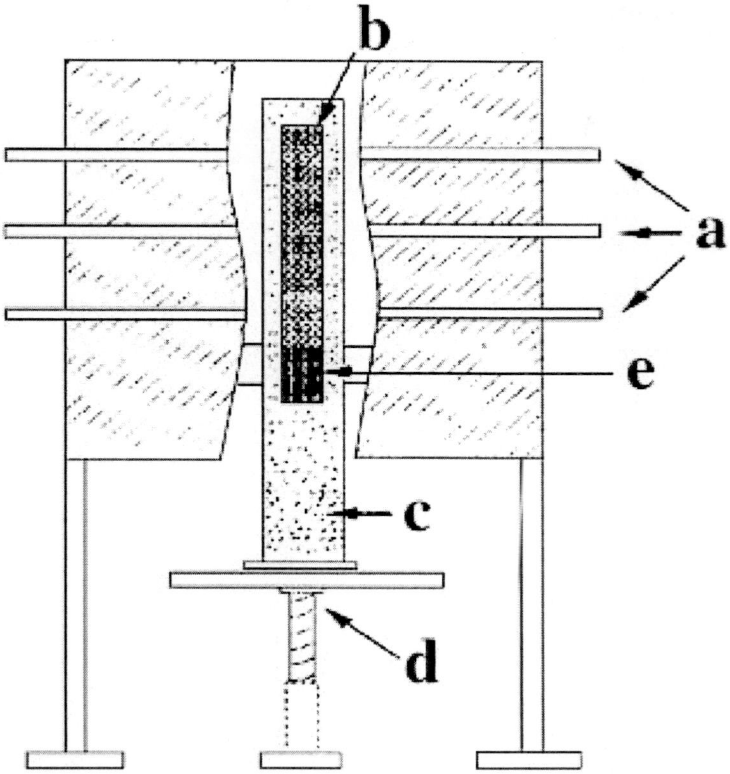


Figure 1. Schematic diagram of the modified Bridgman furnace for the growth of PMN-PT single crystals. (a) Heating equipments, (b) Pt crucible, (c) $Al_2O_3$ powder, (d) Drop-down mechanism, (e) Seed crystal.

## D. Morphology and Growth Mechanism

The process of crystal growth is generally divided into three steps in succession, the dissolution of solute, the formation of growth units and the transportation and combination of growth units on the growth interfaces. The theoretical calculation and experimental results have demonstrated that the growth units in PMN-PT are various $[BO_6]$ coordinate octahedral [34,35]. The structure of PMN-PT single crystals can be regarded as the one made of $[BO_6]$ coordination polyhedra linked by their vertexes. Figure 2 shows the schematic diagram of perovskite structure $ABO_3$ composed of $[BO_6]$ octahedral growth units and $Pb^{2+}$ ions. With the model that the crystal growth units are anionic coordination polyhedrons [34], the relations between the crystal morphology and the building process of these $[BO_6]$ octahedra in the manner of vertex-by-vertex to each face are discussed together with the mechanism of crystal habit of PMN-PT single crystals.

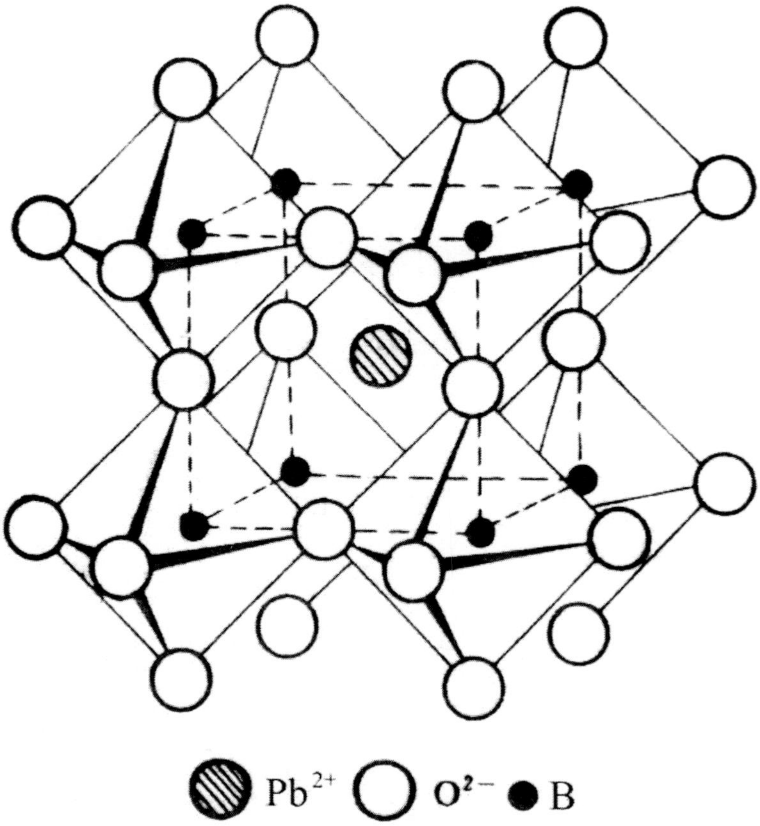


Figure 2. The schematic diagram of perovskite structure $ABO_3$ composed of $[BO_6]$ octahedral growth units and $Pb^{2+}$ ions.

The model considers that the morphology of crystals depends on the orientation of structure units in the crystals and that the growth units in the melts have the same structure units with the crystals themselves. A quenching experiment was taken to investigate the structure of PMN-PT solution from 1310 °C to room temperature. The XRD result show that the structure of PMN-PT melt is local ordering, which is similar to the structure of PMN-PT single crystal, so it is assumed that there are growth units whose structure units are octahedral $[BO_6]$ ($[NbO_6]$, $[MgO_6]$ and $[TiO_6]$) in PMN-PT growth melt [35]. In fact, the surface structures atop the boules, which consist of series of triangle pyramids or the vertexes of cubes, are the locus of the combination of anionic coordination polyhedra $[BO_6]$ on the {111} faces. As for $BaTiO_3$ single crystals with the same structure with PMN-PT, it has been testified indirectly that there are $[TiO_6]$ growth units in the growth solution by the experiments of applying electric fields on the growth solution of $BaTiO_3$ [36]. The calculation on the stability energy of growth units also shows that $[TiO_6]$ could exist in the $BaTiO_3$ solution [36]. Based on above data, we consider that it is very possible that there are $[BO_6]$ growth units, i.e. $[MgO_6]$, $[NbO_6]$ and $[TiO_6]$ octahedra, in PMN-PT melts.

Though the clusters in PMN-PT melt are united and decomposed rapidly, it is believed that the stable clusters exist at the melt. The primal stable clusters are supposed to the octahedral $[NbO_6]$, $[MgO_6]$ and $[TiO_6]$ in melt where the $Pb^{2+}$ cations are linked with these octahedra. The transportation of octahedral $[BO_6]$ ($[NbO_6]$, $[MgO_6]$ and $[TiO_6]$) is more difficult on the growth boundary than $Pb^{2+}$, so the reaction of these octahedra on the growth interface is slower than $Pb^{2+}$ and the transportation and reaction of these octahedral [$NbO_6$], [$MgO_6$] and [$TiO_6$] are the rate-determining processed. The growth mechanism and crystal habit of PMN-PT single crystals could be analyzed through investigating the behavior of the growth units on crystal surface.

With the view of crystal chemistry, the growth rates of different lattice faces are mainly dominated by the structure of growth interfaces, which is constituted by [$BO_6$] growth units and the kinetic effect of unit combination with these interfaces during the crystal growth. The difference in growth rates can be analyzed from the stability of the combination of the growth units and the structure units on the crystal surfaces. When [$BO_6$] octahedra combined with {111}, {110} and {001} faces, they link up with the octahedra on the interfaces at 3, 2, and 1 free vertexes respectively. Thus, the stability energies of combination will decrease successively in this order so as that the growth rates along these faces fall off in turn. As a result, once crystals gain the opportunity of freedom growth, the {001} faces will appear dominantly because it has the slowest growth rate. In fact, under normal conditions, PMN-PT single crystals take normal growth mechanism while [$BO_6$] units combine with kinked faces {111}, and take layer by layer growth mechanism while they combine with flat faces {001}. Hence, the crystals grow the fastest along [111] direction, the most slowly along [001] and have medial growth rate along [110], which leads the dominant appearance of {001} faces. The crystal habits have been reflected in observed phenomena, including the surface morphology in growth process, the surface morphology formed by thermal etching, negative crystal structures and growth steps.

It should be pointed out that the result from above model of anion coordination polyhedron growth units is coincident with that from Periodic Bond Chain (PBC) theory [35]. Periodic Bond Chain (PBC) theory is originally formulated by Hartmann and Perdok to explain or predict the growth habits of small molecule crystals. It can also be employed to investigate crystal growth mechanisms at the molecular level. According to PBC theory, {111}, {110} and {001} faces in PMN-PT single crystals with simple cubic structure are kinked (K), stepped (S) and flat faces (F) respectively and their growth rates decrease successively. It should be noticed that when [$BO_6$] growth units can not be built up for lack of $Pb^{2+}$ arisen from the high volatility of PbO, the negative surface morphology will appear atop the end of boules.

The formation mechanisms of composition fluctuation can also be studied by the model of anion coordination polyhedron growth units. In view of the differences of various [$BO_6$] growth units in stable energy, dimensions and charge valences, Xu and Luo et al. [35] indicated that the split of the growth units with the same structures would be responsible for the micro-fluctuation of crystal compositions and structures through the building process of growth units. In the melt, these different [$BO_6$] units exist in different possibilities, octahedral units [$NbO_6$] and [$TiO_6$] being more favorable growth units than [$MgO_6$]. Due to the fact that the growth interfaces carry some charges, they show a certain choice amongst [$BO_6$] units to be combined, leading to a short range of fluctuation in compositions and structures. The successful growth of pure perovskite PMN-PT single crystals from their melt by the modified Bridgman method can be ascribed to the following accounts: the contribution of [$TiO_6$] to crystal stability, the modulation effect of [$TiO_6$] on micro compositions and structures and the relative stability of [$MgO_6$] compared with [$ZnO_6$].

On the basis of analysis of the growth units, the morphology of as-grown PMN-PT single crystals can be well understood, and the crystal habit can also be well explained. From the patterns of appearance of PMN-PT single crystals varied with growth conditions, the parameters of PMN-PT crystal growth can be optimized to grow PMN-PT single crystals with large size and high quality.

Figure 3. PMN-0.38PT single crystals obtained by a modified Bridgman technique. (a) As-grown crystal boules, (b) After being poled and polished [30].

By the modified Bridgman technique, high-quality PMN-PT single crystals with 50 mm in diameter and about 80 mm in length have been achieved successfully. PMN-0.38PT single crystals are given as an example and shown in Fig. 3 [30]. The resulting crystals are light yellow in color and have smooth and shiny surfaces with the grain structure from the Pt crucible imprinted on the surfaces. There are a few stripes in the surfaces of as-grown crystal boules caused by stress brought about by the lattice deformation during phase transition and by the thermal gradient during crystal growth. The size and quality of the obtained crystals are good enough for various applications, even for optical devices. The poled and polished samples including some crystal ingots and some sectional crystal slices are shown in Fig. 3 (b). The thickness of these crystal slices varies from 0.6 to 1.2 mm and they are transparent.

## III. OPTICAL PROPERTIES

The complex perovskite PMN-PT single crystals are well-known ferroelectrics. They are the solid solutions of PMN with PT and exhibit a nearly vertical MPB near the PT content ($x$) of 0.30 to 0.35 [37,38], dividing the rhombohedral phase and the tetragonal one. For compositions with $x \geq 0.35$, the single crystals have a tetragonal structure with symmetry of P4*mm*, and for those with $x \leq 0.30$, the crystal symmetry is rhombohedral R3*m*. Normal ferroelectric behavior increases as a function of $x$, whereas relaxor behavior decreases with the increasing of $x$ [39].

In 2000, Fu and Cohen pointed out a theoretical consideration of the polarization rotation path under an external electric field [40]. It is a milestone to reveal the origin of high piezoelectric response in the perovskite ferroelectric. Except for explores in theories, many experimental studies have been carried on PMN-PT single crystals for their domain structure, dielectric constant, piezoelectric coefficient, electromechanical coupling factor, field-induced strain response, pyroelectric properties, phase transitions and doping effect et al. over a wide range of temperatures and compositions [1-4,9-11,37-51]. We have also characterized the basic dielectric and piezoelectric properties of PMN-PT single crystals obtained by the modified Bridgman technique and found that they possess excellent electrical properties. For example, for PMN-0.33PT single crystals orientated along (001) face, their electromechanical coupling factors are $k_{33} \sim 94\%$ for the longitudinal mode, $k_{33}' \sim 86\%$ for the beam mode and $k_t \sim 62\%$ for the thickness mode. Their piezoelectric coefficients $d_{33}$ are 2000 ~ 2500 pC/N, and dielectric constants $\varepsilon_r$ are 4500 ~ 5500, and ultrahigh field-induced strain response $S$ up to 1.8% has been achieved [9-14,52]. All the results show that the PMN-PT single crystals grown by the modified Bridgman technique have high qualities and are very promising materials for many applications, such as ultrasonic transducers and strain actuators.

Most oxygen-octahedra ferroelectrics, which exhibit excellent electromechanical properties, also have outstanding optical performance [15,16]. Optical materials with high EO response are widely used in numerous devices, including modulators, EO deflectors, switches and filters, and they are most desired for optical communications. However, in spite of this extensive utilization, the values of the optical properties of PMN-PT single crystals are not perfectly known. It might be attributed to the lack of high-quality single crystals which restricted their further investigations and optical applications. The knowledge of the detailed optical parameters, such as the refractive indices or optical transmission properties, is most desirable, not only for making possible practical usage in optic devices but also to answer very basic questions on the energy level of the crystals and the physical nature of this class of relaxor ferroelectrics.

On the basis of the obtained large-size and high-quality PMN-PT single crystals, we have studied their optical properties systematically. In this chapter, we will report their up-to-date optical properties, including the results of the refractive indices, the optical transmittance, the extinction ratio and the linear EO properties. It has been found that PMN-PT single crystals are also very promising for EO applications.

## A. Sample Preparation

From the grown boules, PMN-PT single crystal samples were oriented using an X-ray diffractometer and cut into the desired dimensions and orientations. Silver paste was painted on the sample surfaces and sintered at 580 °C for 30 minutes. From the anterior discussions, we know that the segregation behavior during the crystal growth results in the variation of PT along the longitudinal direction of a same boule, so we should determine the PT content firstly. For ferroelectric PMN-PT single crystals, the content of PT is very sensitive to $T_m$ (the temperature of the permittivity maximum). With the phase diagram of PMN-PT solid solution system, the relations can be well described as follows [40,52]:

$$T_m = 500x - 10 \tag{2}$$

where $T_m$ is the temperature in Celsius scale and $x$ is the PT content. With the equation, we can select desired samples with different PT contents by measuring their $T_m$.

Then, the selected samples were poled, usually, under an electric field of 10 kV/cm for fifteen minutes near their $T_m$ in silicone oil and then slowly cooled to room temperature with the half of the applied electric field. A Berlincourt-type quasi-static $d_{33}$ meter is used to measure the room temperature piezoelectric constant $d_{33}$ at a frequency of 55 Hz. It is an easy way to judge whether the samples are poled successfully or not. The surfaces for light transmission were polished using alumina and diamond polishing compounds (with decreasing average grit size down to 0.05 μm) to achieve a final optical polish, with specular reflection. It should be mentioned here that the operation temperatures for all processes should be kept no more than 100 °C after the samples being poled, otherwise the PMN-PT single crystal samples will depoled.

## B. Refractive Indices

One of the most important factors determining the use of a new-style material for optical applications is its refractive index $n$. Studies have been reported on the refractive indices of PMN single crystal [53], PMN-0.3PT ceramics [54] and $PbTiO_3$ ceramics [55]. The values are 2.522, 2.598 and 2.668 at 633 nm, respectively. Bing *et al.* further suggested that for the PMN-PT family, the refractive indices increase with PT content [54].

The refractive indices and their dispersion of PMN-0.38PT single crystal were accurately measured with the minimum deviation method. The sample was cut into prism shape with the apex angle of ~ 15°. From the minimum deviation method, the refractive index is determined by using the expression [56]:

$$n = sin\frac{A+D}{2} \Big/ sin\frac{A}{2} \tag{3}$$

where $A$ is the apex angle of the prism and $D$ is the angle of minimum deviation.

The refractive indices of PMN-0.38PT single crystal measured at room temperature are given in Table 1. Curves of $n_o$ and $n_e$ versus wavelength are shown in Fig. 4. We can see that the refractive indices are frequency dependent in visible spectrum and increase dramatically with the decreasing of wavelength. It is well known from simple dispersion theory that in the region of low absorption the refractive index can be given by the modified Sellmeier's equation [39,56]:

$$n_i^2(\lambda) = A_i + \frac{B_i}{\lambda^2 - C_i} - D_i \times \lambda^2 \tag{4}$$

The parameters $A_i$, $B_i$, $C_i$ and $D_i$ can be obtained in terms of least-squares fit and the obtained Sellmeier's equation for PMN-0.38PT single crystal are:

$$\begin{aligned} n_o^2(\lambda) &= 6.2144 + \frac{0.2151}{\lambda^2 - 0.0686} - 0.0004 \times \lambda^2 \\ n_e^2(\lambda) &= 6.0840 + \frac{0.2257}{\lambda^2 - 0.0593} + 0.0499 \times \lambda^2 \end{aligned} \tag{5}$$

Then the refractive indices of PMN-0.38PT single crystal at other wavelength region can be calculated easily from the equation. For example, at 632.8 nm, the calculated data of the refractive indices $n_o$ and $n_e$ are 2.620 and 2.601, respectively.

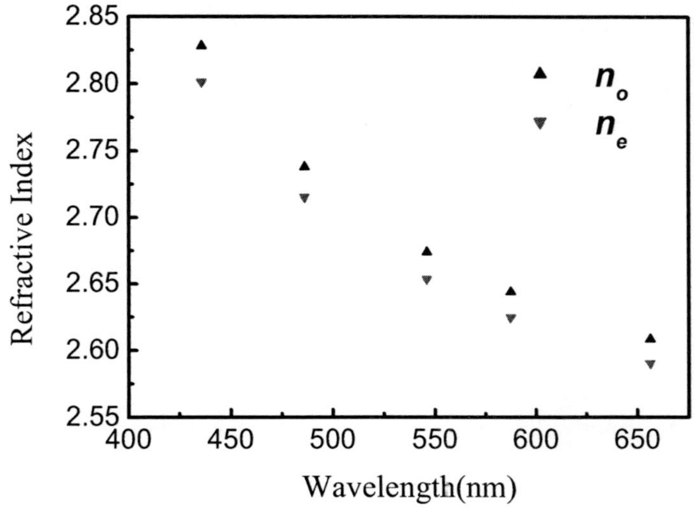


Figure 4. Wavelength dependence of the refractive indices at room temperature of PMN-0.38PT single crystal.

**Table 1. The refractive indices of PMN-0.38PT single crystal at room temperature measured by the minimum deviation method**

| Wavelength (nm) | $n_o$ | $n_e$ |
|---|---|---|
| 435.8 | 2.828 | 2.801 |
| 486.1 | 2.738 | 2.715 |
| 546.1 | 2.674 | 2.654 |
| 587.6 | 2.644 | 2.625 |

The minimum deviation method can be used to measure refractive indices with high accurate, however it is not suitable to apply for the PMN-PT single crystals with $x < 0.35$, i.e. for the PMN-PT in rhombohedral phase and in MPB region. Because there are many domain walls and nanoregions with different orientations which obviously cause multiple scattering and generate large scattering losses, so the PMN-PT single crystals with these compositions become opaque when the thickness is over 1 mm.

A prism coupler (Metricon Model 2010) which can measure refractive index and its anisotropy for bulk sample with high accuracy (The worst case index accuracy is ± 0.001 and resolution is ± 0.0001) is used in our measurement. It usually uses TE-mode (electric field vibrating transverse to the plane of incidence) and TM-mode (magnetic field transverse to and electric field parallel to the plane of incidence) to characterize the anisotropy of refractive indices. But only TE-mode measurement can be used for PMN-PT single crystals because their refractive indices are very large and out of the measurement range of TM-mode [57]. Just with TM-mode, we still can obtain both the values of $n_o$ and $n_e$ by using different incidence plane, respectively, perpendicular to the optic axis for $n_o$ and parallel to the optic axis for $n_e$. The refractive indices of PMN-PT single crystals with different compositions and orientations measured at 633 nm are listed in Table 2. We can see that as the PT content increases, the refractive indices $n_o$ of PMN-PT single crystals increase. The results correspond with the suggestion made by Bing *et al.* and have been observed in PZN-PT single crystals [54]. It is maybe due to the higher refractive index of pure PT (2.668) as compared to that of pure PMN (2.522) [54,58]. By the way, the values of PMN-0.38PT single crystal obtained by the prism coupler ($n_o$ = 2.616 and $n_e$ = 2.600) are slightly lower than the results measured by the minimum deviation method ($n_o$ = 2.620 and $n_e$ = 2.601). It may be caused by the roughness of the surface and the segregation behavior existed in PMN-PT single crystals as anterior discussions.

When a material is applied in devices actually, the influence of the variations of temperature must be considered, so we have also measured the temperature dependence of the refractive indices of PMN-PT single crystals and the results are shown in Table 3. Both the values of $n_o$ and $n_e$ increase slightly with the increasing of temperature. With the measured values of refractive indices at different temperatures, thermo-optic coefficients of PMN-PT single crystals are calculated and the results are listed in the last column of Table 3.

**Table 2. The values of the refractive indices, EO coefficients and the half-wave voltage $V_\pi$ for PMN-PT single crystals at room temperature**

| Composition | Poling direction | $n_o$ | $n_e$ | $r_c$ (pm/V) | $V_\pi$ (V) |
|---|---|---|---|---|---|
| PMN-0.26PT | <111> | 2.584 | 2.595 | 91 | 411 |
| PMN-0.30PT | <111> | 2.590 | 2.598 | 107 | 347 |
| PMN-0.33PT | <111> | 2.598 | 2.602 | 109 | 337 |
| PMN-0.33PT | <001> | 2.599 | 2.562 | 182 | 202 |
| PMN-0.35PT | <001> | 2.602 | 2.588 | 81 | 452 |
| PMN-0.38PT | <001> | 2.616 | 2.600 | 41 | 878 |
| PMN-0.42PT | <001> | 2.622 | 2.603 | 21 | 1708 |

**Table 3. Temperature dependence of the refractive indices for PMN-PT single crystals**

| Sample | 21 °C | 80 °C | $dn/dT$ ($10^{-6}$/K) |
|---|---|---|---|
| PMN-0.35PT, $n_o$, $\lambda$ = 633nm | 2.603 | 2.614 | 186 |
| PMN-0.35PT, $n_e$, $\lambda$ = 633nm | 2.591 | 2.602 | 186 |
| PMN-0.35PT, $n_o$, $\lambda$=1550nm | 2.495 | 2.503 | 136 |
| PMN-0.35PT, $n_e$, $\lambda$=1550nm | 2.487 | 2.497 | 169 |
| PMN-0.38PT, $n_o$, $\lambda$ = 633nm | 2.616 | 2.626 | 169 |
| PMN-0.38PT, $n_e$, $\lambda$ = 633nm | 2.599 | 2.610 | 186 |
| PMN-0.38PT, $n_o$, $\lambda$=1550nm | 2.505 | 2.512 | 119 |
| PMN-0.38PT, $n_e$, $\lambda$=1550nm | 2.497 | 2.506 | 153 |

## C. Transmittance Characterization

Optical transmission properties are not very important for electrical materials, but they are basic and crucial factors for optical materials. Transmission spectra for PMN-PT single crystals were measured at room temperature as a function of wavelength. A Shimadzu UV-2501PC spectrophotometer with the wavelength range from 800 down to 200 nm is employed to obtain the optical absorption edge in near ultraviolet (UV), and a ThermoNicolet Nexus 870 Fourier-transform-infrared (FT-IR) spectrophotometer with the wavelength range from 2.5 - 25 μm is used to investigate infrared optical transmission properties.

Three PMN-PT single crystals with typical compositions, $x$ = 0.29 0.32 and 0.38 for rhombohedral, MPB region and tetragonal sample, respectively, are reported in this chapter. The transmission spectra of the <001>-oriented samples in near UV and visible light region are shown in Fig. 5a. The crystals are all transparent in the visible region and rolls off near 450 nm, and they become completely absorbing around 400 nm, indicating an optical absorption edge in near UV. This is similar to most oxygen-octahedral perovskites [54,59]. The measurements of transmission spectra were extended to 3.1 μm and no absorption was observed. As compared with PMN-0.29PT and PMN-0.32PT single crystals, the transmittance of PMN-0.38PT is much larger, especially in the visible light region and the short wavelength cutoff is much sharper. The optical absorption edge of PMN-0.38PT single crystal lies in a slightly longer wavelength region. That is to say, a remarkable redshift of the intrinsic UV absorption edge is observed between PMN-0.38PT single crystals and the other two compositions. By the way, we have measured a series of <001>-oriented PMN-PT single crystals and found that for the samples of $x = 0.33 \sim 0.40$, the optical transmission properties in near UV and visible light region are almost the same. They have similar optical absorption edge and the maximum transmittance. But the results for other measured compositions ($x = 0.22 \sim 0.33$) are quite different. They have different optical transmission properties and no rule can be found. Even for different samples with the same composition, they still have very different transmittivity each other.

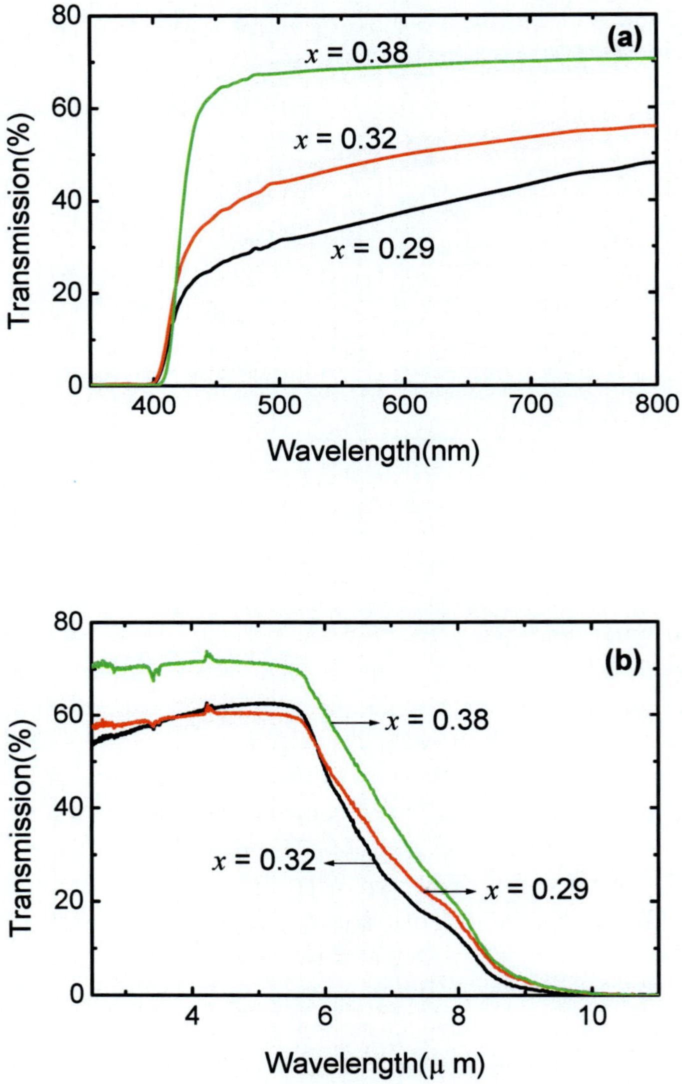


Figure 5. The transmission characteristics of <001>-oriented PMN-PT single crystals in (a) near UV and visible light region, and (b) Infrared region.

The optical transmission properties in Infrared region of the three samples are shown in Fig. 5b. They are all transparent between 2.5 and 5.5 μm and rolls off near 5.5 μm. But the wavelength cutoff is much flatter than the short wavelength cutoff. The samples became completely absorbing around 10 μm. As the transmission spectra in visible light region, the optical transmittance of PMN-0.38PT single crystal is also much larger than that of PMN-0.29PT and PMN-0.32PT, especially in the region of 2.5 to 5.5 μm. So from the optical transmittance characterization in the whole wavelength region, we can conclude that tetragonal PMN-PT single crystals possess the optimal optical transmission properties.

PMN-PT single crystals, like other $ABO_3$-type perovskite compounds, such as PMN, PZN-PT family, have a common $BO_6$ oxygen-octahedra structure. It is already well known via lattice dynamical theories that the static dielectric behavior of ferroelectrics with $BO_6$ octahedron building block is determined primarily by the $BO_6$ octahedra. In the case of optical properties, the $BO_6$ octahedron assumes special significance since the B-site cation *d* orbits and the O-anion 2*p* orbits associated with each octahedron govern the lower lying conduction bands and the upper valence bands. So the $BO_6$ octahedra determine the basic energy level of crystal [54,60]. But the three samples are rhombohedral, MPB and tetragonal, respectively, which causes a slight difference in optical properties. In $Pb(B_1B_2)O_3$ ferroelectric relaxors, the nanoregions, domain structure and order/disorder state of B-site cation structure have influences on the optical transmittance as well as the dielectric and piezoelectric behavior of the crystals. For PMN-PT single crystals, $Mg^{2+}$, $Nb^{5+}$ and $Ti^{4+}$ ions occupy the same crystallographic site in the structure and the crystal has a complex energy distribution. The distortion of lattice, ion vibration and interaction of the ions with different poling direction comprise complex mechanisms to reveal the optical characterizations [30,39,56-61].

The optical transmission spectra from 0.3 to 11 μm of tetragonal PMN-0.38PT single crystal are shown in Fig. 6. As discussed previously, the crystal is transparent between 0.45 and 5.5 μm and becomes completely absorbing around 0.4 μm and 10 μm under poled condition. As compared with unpoled condition which is also shown in Fig. 6, we can conclude that the sample does not go through the phase transition during the poling process because it has a same optical absorption edge before and after poling. But the transmission has increased dramatically after the sample being poled. On the surface and the domain walls of the single crystal, the refractive index has a discontinuity which will reduce the transmission. The unpoled sample obviously has many domain walls and nanoregions with different orientation with respect to the optic axis. When the incident light passes through the crystal, those domain boundaries cause multiple scattering and generate larger scattering losses. For tetragonal PMN-0.38PT single crystal, the spontaneous polarization is along the <001> direction. After the sample being poled along <001> direction, the single domain area enlarges and the losses due to the domain walls reduce substantially.

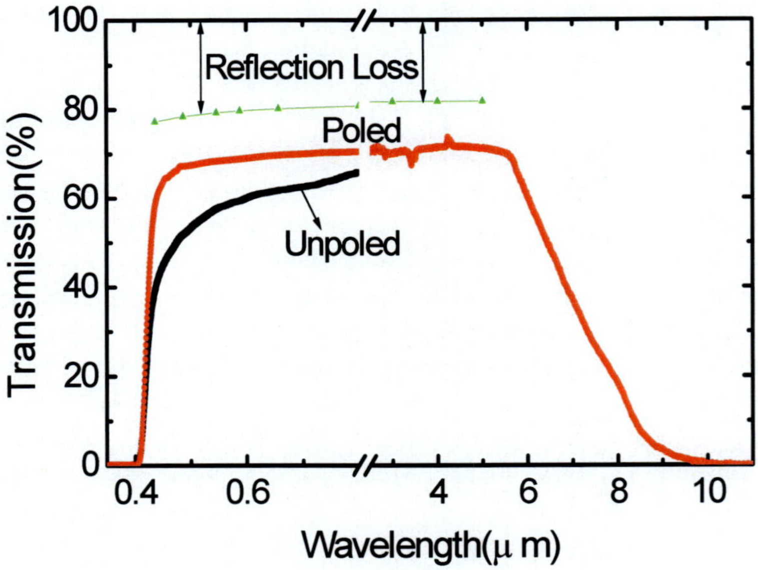


Figure 6. The transmission characteristics of <001>-oriented PMN-0.38PT single crystal. (The wavelength region is 0.3 ~ 11 μm for poled condition and 0.3 ~ 0.8 μm for unpoled condition)

From the transmission characterization, we can see that the optical absorption is very small between 0.45 and 5.5 μm wavelength range for the poled condition. In general, except for absorption losses, optical transmission relates to reflection losses and scattering losses. The reflection losses $R$ of the light at two surfaces can be calculated using the Fresnel expression:

$$R = \frac{(n-1)^2}{(n+1)^2} \tag{6}$$

where $n$ is the refractive index which be calculated from Eq. (5). The calculated values of $R$ are also shown in Fig. 6 and we can see that after considering the reflection losses at both surfaces, the optical transmission for PMN-0.38PT single crystal from 0.45 μm to 5.5 μm wavelength region is quite fine. It indicates that the single crystals can be used as optical materials in a wide wavelength region.

We also measured the extinction ratio of a <001>-cut PMN-0.38PT single crystal with a 4 mm long cube. As shown in Fig. 7, the extinction ratio is about 28 dB at room temperature and changes only slightly from 20 to 70 °C. The high extinction ratio with the independent of temperature is also an important property for use in optical devices.

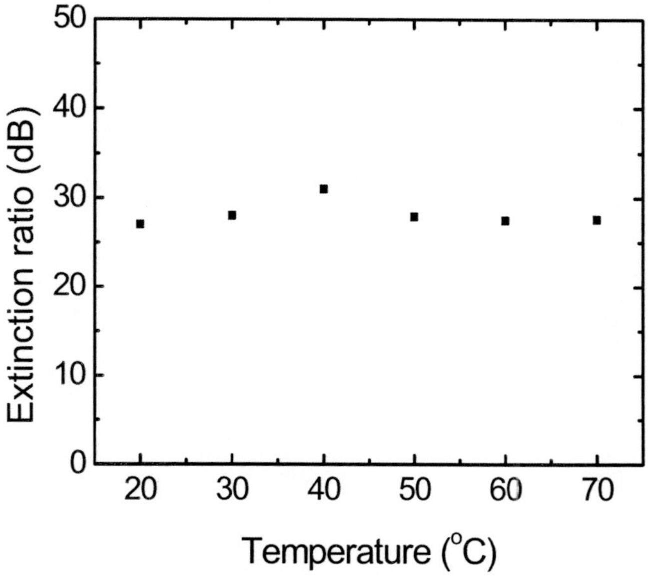


Figure 7. Temperature dependence of the extinction ratio for PMN-0.38PT single crystal with a 4 mm long cube.

## D. Linear EO Properties

The EO effect describes the change in the refractive index ($\Delta n_{ij}$) due to an applied electric field ($E_k$). For poled ferroelectric crystals, we have

$$\Delta n_{ij} = -\frac{1}{2} n_{ij}^3 r_{ijk} E_k \tag{7}$$

where $r_{ijk}$ is the linear EO coefficient. EO effect is used in numerous optical devices, such as optical modulators, EO deflectors, switches and filters. EO measurements are performed with an original and very sensitive method which is based on the Sénarmont compensating setup [57,62]. When the polarization direction of the polarizer is set at $\pi/4$ with respect to the birefringence axis and the electric field is applied along the optic axis of the sample, the effective EO coefficient $r_c$ is obtained, which is a combined coefficient expressed as [56,57,62]:

$$r_c = r_{jjk} - (n_i/n_j)^3 r_{iik} \tag{8}$$

A summary of the effective EO coefficients measured at room temperature is presented in Table 2. The results reveal several interesting features. First, large values of $r_c$ are observed for the PMN-PT single crystals in both the rhombohedral and tetragonal phases. They are all higher than that of widely used EO single crystal $LiNbO_3$, whose $r_c$ is 19.9 pm/V [63]. These excellent EO properties, as well as being able to obtain high quality single crystals with large size, make PMN-PT single crystals a promising candidate for EO material. Second, large $r_c$ is obtained for the compositions near the MPB, as shown in Fig. 8. This optical property may be attributed to the phase transition between the tetragonal and rhombohedral phases.. The phase transition due to the electric field and stress is associated with large changes in the crystal lattice constants and the refractive index. Third, for the same composition near the MPB (e.g., PMN-0.33PT single crystal), the $r_c$ value of the crystal poled along the <001> direction is much higher than that of the crystal poled along the <111> direction, as shown in Table 2 and Fig. 8. The result is similar with that of the PZN-PT single crystals [64] and is testified to cause by their spontaneous polarization and dielectric constants as discussed below.

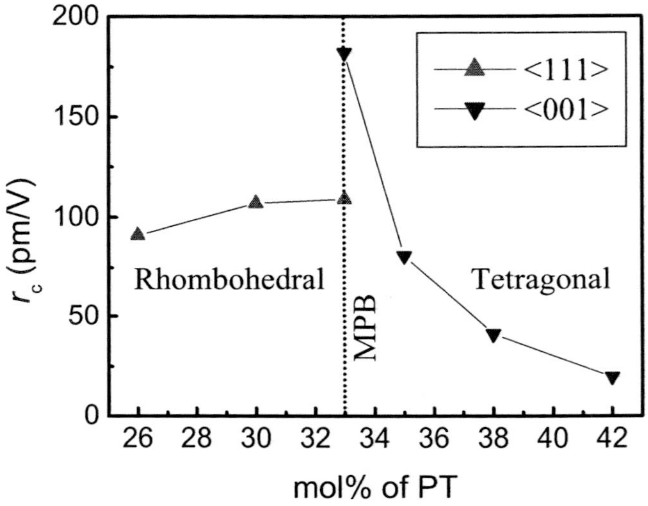


Figure 8. Composition dependence of the EO coefficients of PMN-PT single crystals at room temperature.

It should be mentioned here that the qualitative aging experiments have been performed and it has been found that even the poled single crystals are used for various other measurements at room temperature, the EO coefficients are nearly the same when measured after several months. Therefore, the EO properties reported here are the stable material properties for PMN-PT single crystals.

We now discuss the temperature dependence of these EO coefficients, which is also important for applications in optical devices. As shown in Fig. 9a, the EO coefficients are nearly invariant for tetragonal PMN-0.35PT and PMN-0.38PT single crystals with temperature increases from 20 to 80 °C, while increase with temperature for PMN-0.33PT single crystal. The temperature dependent property may also be related with their dielectric constant and spontaneous polarization. For ferroelectric single crystals in tetragonal phase, the measured linear EO coefficients can be related to the polarization-related quadratic EO coefficients $g_{11}$ and $g_{12}$ as follows [15,16,57]:

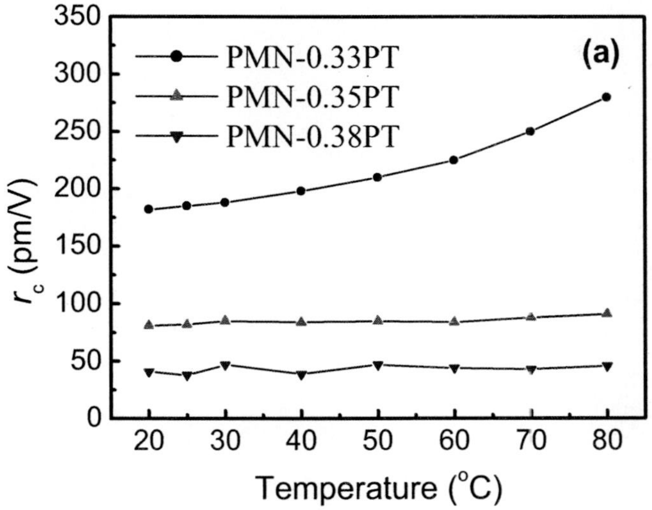


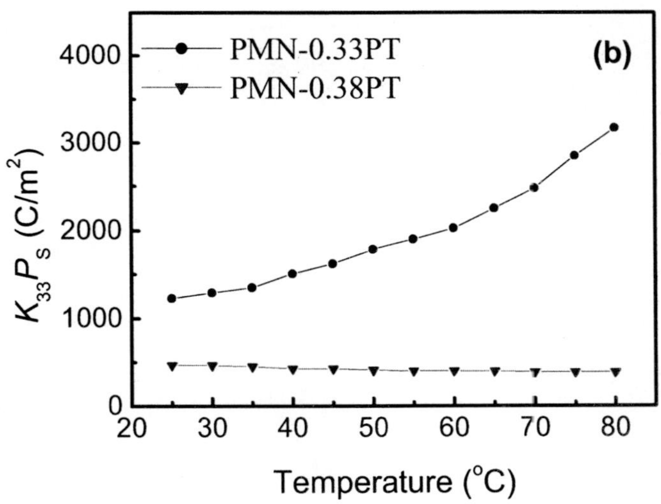


Figure 9. Temperature dependence of (a) the EO coefficients and (b) the dielectric constant times the spontaneous polarization ($K_{33}P_S$) of <001>-oriented PMN-PT single crystals.

$$r_{33} = 2K_{33}\varepsilon_0 P_S g_{11} \text{ and } r_{13} = 2K_{33}\varepsilon_0 P_S g_{12} \tag{9}$$

where $P_S$ and $K_{33}$ are the spontaneous polarization and dielectric constant, respectively, and $\varepsilon_0$ = 8.85 pF/m is the vacuum permittivity. From Eqs. (8) and (9), we can obtain:

$$r_c = 2K_{33}\varepsilon_0 P_S\left[g_{11} - \left(\frac{n_o}{n_e}\right)^3 g_{12}\right] \tag{10}$$

As $g$ coefficients are nearly invariant with temperature and frequency [15,57,64], the changes of the linear EO coefficients $r_c$ should be determined by the dielectric constant times the spontaneous polarization ($K_{33}P_S$). From the values of $K_{33}$ and $P_S$ [57,65], the temperature dependences of $K_{33}P_S$ for PMN-0.33PT and PMN-0.38PT single crystals are obtained and plotted in Fig. 9b. Indeed, the variations are very similar to those of the temperature dependence of $r_c$ values. Moreover, the large values of $r_c$ appear in the compositions near the MPB and their composition and orientation dependence both result from the values of $K_{33}P_S$.

The half-wave voltage $V_\pi$ is an important figure of merit for EO modulator. It is often used for selecting bulk EO materials and is defined by [62]:

$$V_\pi = \lambda / \left(n_e^3 \cdot r_c\right) \tag{11}$$

$V_\pi$ corresponds to the voltage needed to obtain a phase retardation of $\pi$ from a bulk sample which is independent of its dimensions. In order to obtain a large efficiency for EO modulation applications, $V_\pi$ should be as small as possible. As shown in Table 2, PMN-PT single crystals also have much lower $V_\pi$ compared with that of $LiNbO_3$ single crystal (2800 V) [66]. Higher EO coefficients and lower $V_\pi$ enable operation at lower voltages and smaller device dimensions. Thus, PMN-PT single crystals are of more interest for the development of EO devices instead of $LiNbO_3$ single crystal.

## IV. Conclusion

It is very important to study the feasibility of the growth of PT-based relaxor single crystals if one will develop a method to grow high-quality and large-size single crystals in a production-viable and cost-effective way. For the growth of PMN-PT single crystals, the sticking point is to avoid formation of pyrochlores and prevent evaporation of PbO. PMN-PT single crystals can be grown directly from melt based on their stability of perovskite phase at high temperature, so the modified Bridgman technique with the B-site precursor synthesis method, which has been verified, is a feasible method to grow PMN-PT single crystals in a cost-effective and production-viable way.

Some modifications are necessary and very important to obtain large-size and high-quality PMN-PT single crystals. First, sealed platinum crucibles with appropriate thickness must be used to prevent the evaporation of PbO. Second, high-purity starting materials, usually more than 99.99%, are used to avoid the forming of Pt-metal alloys with lower melting point which will erode Pt crucibles. Third, in order to avoid melt inhomogeneity problems, the component oxides of raw materials are precalcined by the B-site precursor synthesis method and it is also favorable for the stability of pure perovskite phase and restraining the pyrochlore phase during crystal growth. Fourth, seed crystals are usually used during crystal growth and they have a key effect on restraining spontaneous nucleation and parasitic growth. By the modified Bridgman technique, high-quality PMN-PT single crystals with 50 mm in diameter and about 80 mm in length have been obtained successfully.

The optical properties of PMN-PT single crystals were systematically studied. From the results of the refractive indices, optical transmission properties and the extinction ratio, we can conclude that the obtained PMN-PT single crystals can be used as optical materials in a wide wavelength region. It has also been found that the PMN-PT single crystals possess excellent linear EO properties. The values of linear EO coefficients are much larger than that of widely used EO single crystal $LiNbO_3$. The calculated half-wave voltages are also much lower, which enables operation at lower voltages and smaller device dimensions. These excellent EO properties, as well as being able to obtain large-size and high-quality single crystals, make PMN-PT single crystals a very promising candidate for EO modulation applications.

## 5. Acknowledgments

The authors thank Mr. Haiqing Xu, Mr. Tianhou He and Mr. Di Li sincerely for their work in single crystal growth. This work is supported by the National Natural Science Foundation of China (Grant No. 50272075), the High Technology and Development Project of the People's Republic of China (Grant No. 2002AA325130), Hong Kong Research Grants Council (Grant No. PolyU 5193/00P) and the Centre for Smart Materials of the Hong Kong Polytechnic University.

## 6. References

[1] S.-E. Park; T.R. Shrout *J. Appl. Phys.* 1997, *82*, 1804.
[2] R.F. Service *Science* 1997, *275*, 1878.
[3] S.-E. Park; T.R. Shrout *IEEE Trans. Ultrason. Ferroelectr. & Freq. Control.* 1997, *44*, 1140.
[4] Y. Yamashita; Y. Hosono; K. Harada; N. Yasuda *IEEE Trans. Ultrason. Ferroelectr. & Freq. Control.* 2002, *49*, 184.
[5] M. Dong; Z.-G. Ye *J. Crystal Growth* 2000, *209*, 81.
[6] G.A. Smolenskii; V.A. Isupov; V.A. Agranovskaya; S.N. Popov *Soviet Phys.-Solid State* 1961, *2*, 2584.
[7] E.V. Colla; N.K. Yushin; D. Viehland *J. Appl. Phys.* 1998, *83*, 3298.
[8] X. Jiang; F. Tang; J.T. Wang; T.-P. Chen *Physica C* 2001, *364-365*, 678.
[9] H. Luo; G. Xu; P. Wang; Z. Yin *Ferroelectrics* 1999, *231*, 97.
[10] H. Luo; G. Xu; H. Xu; P. Wang; Z. Yin *Jpn. J. Appl. Phys.* 2000, *39*, 5581.
[11] Z. Yin; H. Luo; P. Wang; G. Xu *Ferroelectrics* 1999, *229*, 207.
[12] G. Xu; H. Luo; P. Wang; H. Xu; Z. Yin *Chin. Sci. Bull.* 2000, *45*, 491.
[13] G. Xu; H. Luo; H. Xu; Z. Qi; Z. Yin *Jpn. J. Appl. Phys.* 2001, *40*, L396.
[14] G. Xu; H. Luo; H. Xu; Z. Qi; P. Wang; W. Zhong; Z. Yin *J. Crystal Growth* 2001, *222*, 202.
[15] M. DiDomenico, Jr.; S.H. Wemple *J. Appl. Phys.* 1969, *40*, 720.

[16] S.H. Wemple; M. DiDomenico, Jr. *J. Appl. Phys.* 1969, *40*, 735.
[17] A.S. Bhalla; R. Guo; L.E. Cross; G. Burns; F.H. Dacol; R.R. Neurgaonkar *J. Appl. Phys.* 1992, *71*, 5591.
[18] A.S. Bhalla; R. Guo; L.E. Cross; G. Burns; F.H. Dacol; R.R. Neurgaonkar *Phys. Rev. B* 1987, *36*, 2030.
[19] B. Fang; H. Xu; T. He; H. Luo; Z. Yin *J. Am. Ceram. Soc.* 2004, *71*, 991.
[20] S. Shimanuki; S. Saito; Y. Yamashita *Jpn. J. Appl. Phys.* 1998, *37*, 3382.
[21] K. Harada; Y. Hosono; S. Saitoh; Y. Yamashita *Jpn. J. Appl. Phys.* 2000, *39*, 3117.
[22] B. Fang; H. Xu; T. He; H. Luo; Z. Yin *J. Crystal Growth* 2002, *244*, 318.
[23] B. Fang; H. Luo; H. Xu; T. He; Z. Yin *Jpn. J. Appl. Phys.*2001, *40*, L1379.
[24] G. Xu; H. Luo; Y. Guo; Y. Gao; H. Xu; Z. Qi; W. Zhong; Z. Yin *Solid State Commun.* 2001, *120*, 321.
[25] B. Fang; H. Xu; Y. Wu; H. Luo; Z. Yin *Mater. Lett.* 2002, *52*, 423.
[26] Y. Guo; H. Xu; H. Luo; G. Xu; Z. Yin *J. Crystal Growth* 2001, *226*, 111.
[27] Y. Gao; H. Xu; Y. Wu; T. He; G. Xu; H. Luo *Jpn. J. Appl. Phys.*2001, *40*, 4998.
[28] Y. Guo; H. Luo; T. He; Z. Yin *Solid State Commun.* 2002, 1*23*, 417.
[29] Y. Guo; H. Luo; X. Pan; Z. Yin *Mater. Res. Bull.* 2003, *38*, 857.
[30] X. Wan; J. Wang; H.L.W. Chan; C.L. Choy; H. Luo; Z. Yin *J. Crystal Growth* 2004, *263*, 251.
[31] David R. Lide, ed. *CRC Handbook of Chemistry and Physics*; 71st Edition; CRC Press: Boca Raton, FL, 1990; pp 4-73,77.
[32] S.L. Swartz; T.R. Shrout *Mater. Res. Bull.* 1982, *17*, 1245.
[33] H. Cao; B. Fang; H. Xu; H. Luo *J. Inorg. Mater.* 2003, *18*, 50.
[34] W. Zhong; G. Liu; E. Shi *Science in China Series B* 1994, *37*, 1288.
[35] G. Xu; H. Luo; W. Zhong; Z. Yin; H. Xu; Z. Qi; K. Liu *Science in China Series E* 1999, *29*, 405.
[36] E. Shi; R. Yuan; C. Xia *Acta Phys. Sin.* 1997, *46*, 1.
[37] A.K. Singh; D. Pandey *J. Phys.: Condens. Matter* 2001, *13*, L931.
[38] Y. Guo; H. Luo; D. Ling; H. Xu; T. He; Z. Yin *J. Phys.: Condens. Matter* 2003, *15*, L1.
[39] X. Wan; R.K. Zheng; H.L.W. Chan; C.L. Choy; X. Zhao; H. Luo *J. Appl. Phys.* 2004, *96*, 6574.
[40] H. Fu; R.E. Cohen *Nature* 2000, *281*, 403.
[41] Z.-G. Ye; B. Noheda; M. Dong; D.E. Cox; G. Shirane *Phys. Rev. B* 2001, *64*, 184114.
[42] D. Viehland; J. Powers; L.E. Cross; J.F. Li *Appl. Phys. Lett.* 2001, *78*, 3508.
[43] M.C. Shin; S.J. Chung; S.-G. Lee; R.S. Feigelson *J. Crystal Growth* 2004, *263*, 412.
[44] X. Zhao; J. Wang; H.L.W. Chan; C.L. Choy; H. Luo *J. Phys.: Condens. Matter* 2003, *15*, 6899.
[45] X. Zhao; B. Fang; H. Cao; Y. Guo; H. Luo *Mater. Sci. Eng. B* 2002, *96*, 254.
[46] Y. Guo; H. Luo; K. Chen; H. Xu; X. Zhang; Z. Yin *J. Appl. Phys.* 2002, *92*, 6134.
[47] H. Cao; B. Fang; H. Luo; Y. Sun; J. Guo *Ceram. Int.* 2003, *29*, 145.
[48] H. Cao; B. Fang; H. Xu; H. Luo *Mater. Res. Bull.* 2002, *37*, 2135.
[49] X. Wan; X. Tang; J. Wang; H.L.W. Chan; C.L. Choy; H. Luo *Appl. Phys. Lett.* 2004, *83*, 4711.
[50] L.E. Cross *Ferroelectrics* 1987, *76*, 241.
[51] Z.-G. Ye *Key Eng. Mater.* 1998, *155-156*, 81.
[52] Z. Feng; H. Luo; Y. Guo; T. He; H. Xu *Solid State Commun.* 2003, 1*26*, 347.

[53] D. McHenry; J. Giniewicz; T. Shrout; S. Jang; A. Bhalla *Ferroelectrics* 1990, *102*, 160.
[54] Y. Bing; R. Guo; A. Bhalla *Ferroelectrics* 1987, *242*, 1.
[55] Landolt-Bornstein *Ferroelectrics and Related Substances*; Springer: Berlin, 1981; pp16.
[56] X. Wan; H. Xu; T. He; D. Lin; H. Luo *J. Appl. Phys.* 2003, *93*, 4766.
[57] X. Wan; H. Luo; X. Zhao; D. Wang; H.L.W. Chan; C.L. Choy *Appl. Phys. Lett.* 2004, *85*, 5233.
[58] X. Wan; H.L.W. Chan; C.L. Choy; X. Zhao; H. Luo *J. Appl. Phys.* 2004, *96*, 1387.
[59] X. Wan; H. Luo; J. Wang; H.L.W. Chan; C.L. Choy *Solid State Commun.* 2004, *129*, 401.
[60] Shoichiro Nomura; Hideo Arima; Fumiko Kojima *Jpn. J. Appl. Phys.* 1973, *12*, 531.
[61] X. Wan; T. He; D. Lin; H. Xu; H. Luo *Acta Phys. Sin.* 2003, *52*, 2319.
[62] M. Aillerie; N. Théofanous; M.D. Fontana *Appl. Phys. B: Lasers Opt.* 2000, *70*, 317.
[63] M. Aillerie; M.D. Fontana; F. Abdi; C.C.-Nedelec; N. Théofanous; G. Alexakis *J. Appl. Phys.* 1989, *65*, 2406.
[64] Y. Barad; Y. Lu; Z.-Y. Cheng; S.-E. Park; Q. Zhang *Appl. Phys. Lett.* 2000, *77*, 1247.
[65] X. Zhao; J. Wang; K.-H. Chew; H.L.W. Chan; C. Choy; Z. Yin; H. Luo *Mater. Lett.* 2004, 58, 2053.
[66] J.P. Salvestrini; M.D. Fontana; M. Aillerie; Z. Czapla *Appl. Phys. Lett.* 1994, *64*, 1920.

In: Trends in Crystal Growth Research
Editor: George V. Karas, pp. 71-99
ISBN 1-59454-541-3


*Chapter 4*

# ON THE FORMATION OF TIC CRYSTALS DURING PROCESSING OF NITI SHAPE MEMORY ALLOYS

***Zhonghua Zhang[1,2*], Jan Frenzel[1], Christoph Somsen[1], Josef Pesicka[3], Klaus Neuking[1] and Gunther Eggeler[1]***

[1]Institut für Werkstoffe, Ruhr-Universität Bochum, D-44780 Bochum, Germany
[2]Institute of Material Science and Engineering, Ocean University of China, Yushan Road 5, Qingdao 266003, P.R. China
[3]Charles University, Faculty of Mathematics and Physics, Department of Metal Physics, 12116 Prague, Czech Republic

## ABSTRACT

NiTi shape memory alloys (SMAs) are presently successful in engineering and medical technology. They combine good structural and functional properties with good corrosion resistance and biocompatibility. It is well known that shape memory properties of NiTi SMAs strongly depend on the processing technique and the associated microstructure. But the elementary processes which govern the chemical composition of the alloy during melting and the formation of microstructure during solidification are not yet fully understood. In the present study we investigate the formation of titanium carbide (TiC, NaCl-type B1 crystal structure) which forms during an important ingot metallurgy processing route, where NiTi SMAs are produced from pure elements using vacuum induction melting (VIM) in graphite crucibles. This procedure causes some carbon contamination of the alloy which can be tolerated up to a certain limit (e.g. industrial standard: 0.05 wt.%). We use optical microscopy (OM), X-ray diffraction (XRD), scanning electron microscopy (SEM) with energy dispersive X-ray (EDX) spectrometry, and transmission electron microscopy (TEM) to study the morphology of the carbides and to establish how microstructures form and whether or not there are specific orientation relationships between the NiTi parent phase (B2) and TiC. The solidification behaviour of carbon-containing NiTi melts is discussed on the basis of a quasi binary NiTi-TiC phase diagram. Only eutectic TiC phase was found for NiTi melts with low

* E-mail: zhonghua.zhang@rub.de, zh_zhang@sdu.edu.cn, Fax: 0049-(0)234-3214235

carbon contents. Small (regular octahedral and dendritic) and large (irregular) primary carbides appear in the microstructure of NiTi alloys as the carbon content increases. Our study also shows that <100> is a preferred growth direction of eutectic and small primary TiC carbides during solidification. In addition, TEM results suggest that there exists a specific orientation relationship between the eutectic TiC phase and the B2 matrix.

**Key words:** NiTi shape memory alloys (SMAs), titanium carbide (TiC), scanning electron microscopy (SEM), transmission electron microscopy (TEM)

## 1. Introduction

In the present study we investigate the formation of titanium carbide (TiC) during metallurgy processing of NiTi shape memory alloys (SMAs) in graphite crucibles using vacuum induction melting (VIM). TiC has a cubic NaCl-type (B1) crystal structure (space group Fm3m). It exhibits a wide stability range extending from $TiC_{0.49}$ to $TiC_{1.0}$ [1,2]. Titanium carbide is a high-temperature structural material with very high melting point (3340 K), extreme hardness, low density (4.93 $g \cdot cm^{-3}$), high mechanical stiffness, excellent resistance to oxidation and corrosion, and high thermal and electrical conductivity almost independent of temperature [3-8]. Therefore, it is used in a variety of advanced applications including cutting tools and wear-resistant coatings [5]. It can be fabricated in continuous or patterned thin films for interconnects in modern integrated electronic circuits, and due to its high thermal stability it is particularly used in electronic circuits which are exposed to elevated temperatures [9]. Considerable attention has been paid to the development of *in situ* metal matrix composites (MMCs) reinforced by TiC [10-14]. TiC single crystals have been used as cold field emitters in high-resolution scanning electron microscopes and other micro probe instruments [15]. In addition to the thermal stability they show low values of the work function and are therefore preferred to other materials [15]. TiC can be prepared by chemical vapour deposition (CVD) [16-17], floating zone methods [18-19], and flux/high temperature solution (HTS) methods using Al, Mg, Ni, Co and Fe as solvents [3-6, 20-22]. Recently, TiC nanowires and nano-hollow polyhedrons were prepared using a vapour-liquid-solid (VLS) method [7] and a chemical metathesis route [8], respectively. NiTi SMAs combine good structural properties with excellent shape memory (SM) effects [23-27]. Moreover, they exhibit good corrosion resistance, biocompatibility and a high specific electric resistance (which is beneficial for direct current heating) [25,26,28]. In addition, the shape memory effects of NiTi SMAs (mechanical memory: pseudo elasticity, PE; thermal memory: one way effect, 1WE) can be exploited many times before functional or structural fatigue limits their exploitable service lives [29-33]. Emerging applications of NiTi SMAs include micro-electro-mechanical systems (MEMS) [34,35], bio-medical devices and implants [36-38], and others [39].

So far, the formation of TiC in NiTi SMAs during graphite crucible VIM processing has not received much attention. As compared to vacuum arc melting (VAM), VIM processing is the most prominent laboratory method to produce NiTi SMAs [40-49]. It yields NiTi alloys of good chemical homogeneity and graphite crucibles show a good performance because they are easy to be handled and are available at reasonable prices. VIM processing in graphite crucibles is known to increase the concentration of carbon in the alloy. This carbon forms TiC which affects the Ni concentration of the alloy and the Ni concentration is a critical parameter for adjusting functional and structural SM properties. It is well known that an increase of 0.1 at.% in Ni results in a 10 K decrease in phase transition temperatures [23,28,50], as shown in Fig. 1. Moreover, the morphology, size and distribution of the TiC phase may have an effect on mechanical properties (tensile strength, elongation, fatigue resistance) of NiTi SMAs. TiC particles, especially large primary carbides, could act as crack initiation during cyclic loading. While these are well appreciated very little is known about how the TiC phase forms and this is what we investigate in the present study.

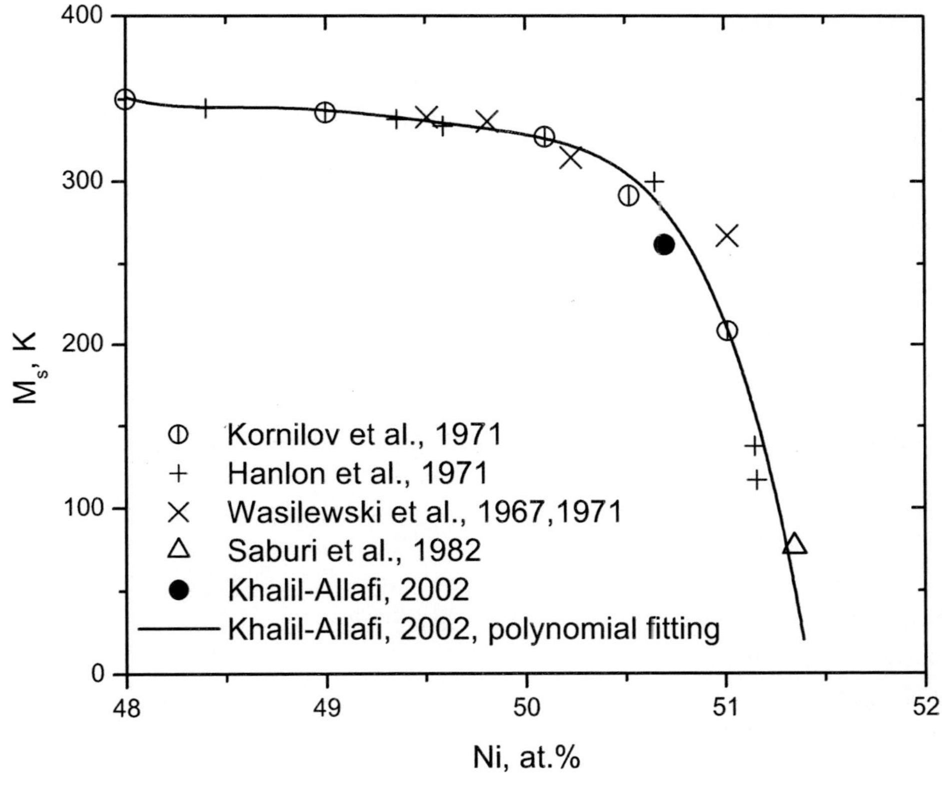


Figure 1. Dependence of $M_s$ temperature on the Ni concentration of solution annealed NiTi alloys [23,28,50].

## 2. Experimental

### 2.1. Materials

In the present study, NiTi (Ni:Ti, 50.7:49.3 in at.%) alloys were prepared from pure elements, Ni pellets and Ti rods. High purity Ni pellets (Ni: >99.98 wt.%, C: 0.012 wt.%, Fe: 0.005 wt.%, Cu: 0.0002 wt.%, S: 0.0004 wt.%) of an average size of 9 mm were purchased from Ampere GmbH, Frankfurt am Main, Germany. Ti rods (Ti: >99.7 wt.%, C: 0.01 wt.%, O: 0.0281 wt.%, Fe: 0.02 wt.%, N: 0.011 wt.%, H: 0.0007 wt.%) of 20 mm in diameter were purchased from E. Wagener GmbH, Stuttgart, Germany. Special care was taken to clean the

Ti surfaces by grinding with 80 mesh SiC emery paper. After grinding, the Ti rods were cleaned using an iterative procedure involving acetone rinsing and alcohol exposure in an ultrasonic bath; the rods were dried prior to melting. Previous work has shown that the quality of graphite crucibles strongly affects the alloy composition [51]; in the present work, graphite crucibles were purchased from Jürgen Dopf Handelsvertretungen, Pforzheim, Germany. A schematic of the crucible is shown in Fig. 2a and its physical properties were given elsewhere [51,52].

## 2.2. Melting

Melting was performed using a Leybold Hereaus IS 1/III induction furnace operating at 5-20 kW under a high purity Argon atmosphere (99.998 vol.%). Prior to melting, the furnace was evacuated to 3 mbar and then filled with Ar up to a pressure of 500 mbar. This flushing was repeated five times. The charges were filled into the crucible using the cladding procedure proposed in [52]. The temperature of NiTi alloys in graphite crucibles was measured by a two-colour pyrometer (Impac ISQ 10-LO ) which was calibrated against three pure element (Si, Fe and Ti) melting points and one NiTi (with 50.7 at.% Ni) melting point (taken from the binary phase diagram). The measuring error of the pyrometer is within ±10 K.

For the investigation of the carbon enrichment in NiTi melts, melting was performed at three temperatures, 1673 (1), 1773 (2) and 1888 K (3) and probes (P1-1 to P1-5, P2-1 to P2-5 and P3-1 to P3-5, Table 1) were taken after reaction times between 60 and 2940 s using small moulds (Fig. 2b). The insert of Fig. 2b shows an example of a resulting ingot. Finally the melt solidified in the graphite crucibles and this yielded three additional ingots (C1, C2 and C3, Table 1). All details of melting procedure discussed so far have been given elsewhere [53]. In addition an effort was made to study the solidification behaviour of NiTi melts in larger Y-shaped moulds (lower cooling rates), Fig. 2c. For this purpose two additional experiments were performed at 1823 K, referred to as LM1 and LM2 in Table 1. These two moulds were prepared in the same way as was described for the small moulds in [53].

## 2.3. Chemical and Microstructural Analysis

From the small ingots (P-series), disks were cut from the lower part of the ingot. In case of the melts which solidified in the crucibles (C-series), specimens were cut along the longitudinal axis of the crucible. Arbitrarily oriented specimens were taken from the two LM-series ingots. The carbon contents of P- and LM-series ingots were determined using the combustion infrared absorption method [54]. From each ingot, a 3-5 g specimen was cut, ground to a final mesh size of 500 and cleaned using the same procedure as has been described for the Ti rods (see above) and chemically analyzed. The measurements were subcontracted to Revierlabor, Essen, Germany.

**Table 1. Series of vacuum induction melting experiments with graphite crucibles performed in the present study. Temperature of the melt, reaction time, resulting carbon concentration and cooling rate (see Table 2) are listed. Moreover, a short description of resulting microstructures is provided (E-eutectic TiC, SP-small primary TiC, LP-large primary TiC, NiTi-the mtrix (B2+B19') and OX-oxide $Ti_4Ni_2O_x$)**

| Ingots | | Parameters | | | |
|---|---|---|---|---|---|
| Set | Sample No. | *t*, s | *c*(*t*) wt.% | CR | Microstructure/Phase |
| 1673 K | P1-1 | 60 | 0.026 | CR1 | E+NiTi |
| | P1-2 | 240 | 0.039 | | E+NiTi |
| | P1-3 | 540 | 0.048 | | E+SP+NiTi |
| | P1-4 | 1500 | 0.061 | | E+SP+LP+NiTi |
| | P1-5 | 2940 | 0.079 | | E+SP+LP+NiTi |
| | C1 | 2940 | 0.079 | CR3 | E+SP+LP+NiTi+OX |
| 1773 K | P2-1 | 60 | 0.036 | CR1 | E+NiTi |
| | P2-2 | 240 | 0.064 | | E+SP+NiTi |
| | P2-3 | 540 | 0.083 | | E+SP+NiTi |
| | P2-4 | 1500 | 0.11 | | E+SP+LP+NiTi |
| | P2-5 | 2940 | 0.14 | | E+SP+LP+NiTi |
| | C2 | 2940 | 0.14 | CR3 | E+SP+LP+NiTi+OX |
| 1888 K | P3-1 | 60 | 0.062 | CR1 | E+SP+NiTi |
| | P3-2 | 240 | 0.13 | | E+SP+NiTi |
| | P3-3 | 540 | 0.16 | | E+SP+LP+NiTi |
| | P3-4 | 1500 | 0.19 | | E+SP+LP+NiTi |
| | P3-5 | 2940 | 0.25 | | E+SP+LP+NiTi |
| | C3 | 2940 | 0.25 | CR3 | E+SP+LP+NiTi+OX |
| 1823 K | LM1 | 10 s | 0.02 | CR2 | E+NiTi |
| | LM2 | 10 s | 0.071 | | E+SP+NiTi+($Ni_4Ti_3$) |

Specimens from each ingot were assessed using optical microscopy (OM), X-ray diffraction (XRD), scanning electron microscopy (SEM) with energy dispersive X-ray (EDX) spectrometry, and transmission electron microscopy (TEM). OM specimens were prepared by grinding and mechanically polishing and investigated using a Leitz Aristomat microscope. Some of the OM specimens were chemically analyzed in the SEM (LEO 1530 VP) by EDX. SEM specimens received an additional surface treatment consisting of grinding to a final mesh size of 1000 and then electropolishing for 7.5 minutes using an electrolyte containing 78.6 vol.% acetic acid and 21.4 vol.% perchloric acid at 10 V (against a Ni cathode) and 293 K [55]. For TEM observations (Philips CM 20), small disks with a diameter of 3 mm and thickness of 0.1-0.15 mm were prepared by grinding. These disks were thinned by a double jet electropolishing technique (Struers TenuPol-5) using an electrolyte containing 95 vol.% acetic acid and 5 vol.% perchloric acid at 19.5 V and 288 K. TEM micrographs were taken to document microstructures and electron diffraction was performed for identification of phases, growth directions and orientation relationships. Additional phase information was obtained using X-ray diffraction (XRD): XRD studies were performed using a Philips X'Pert diffractometer operating with Cu $K_\alpha$ radiation at 40 kV and 40 mA.

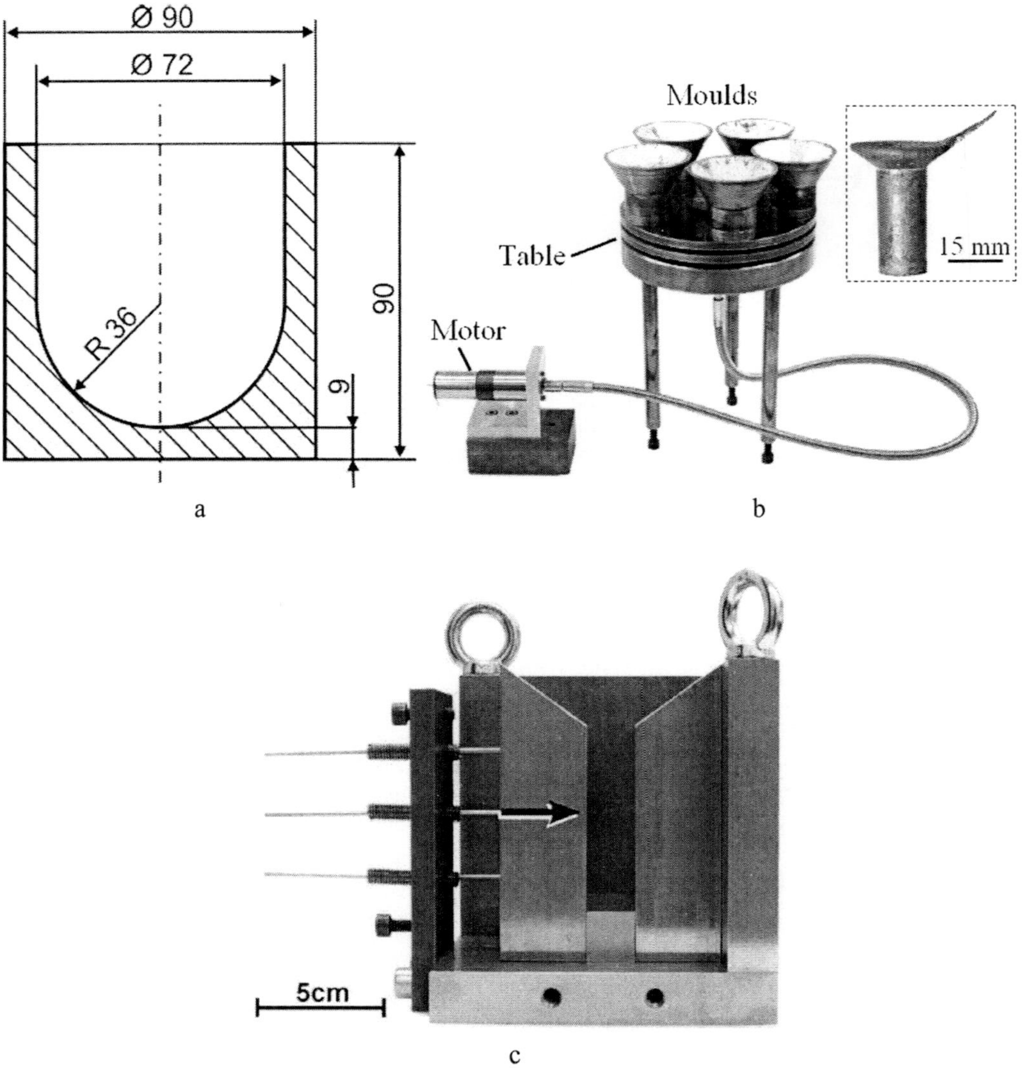


Figure 2. Moulds used for casting of NiTi alloys. (a) and (b) have been described previously [53]. (c) Additional mould used in the present study.

## 2.4. Cooling Rates

In the present work, cooling rates for different solidification conditions were documented. As an example, the cooling curve for the melt which solidified from a melting temperature of 1773 K (solidification in crucible) is shown in Fig. 3. The cooling rate was determined from the gradient of the cooling curve. For the cooling curve shown in Fig. 3 a representative cooling rate of 1 $K \cdot s^{-1}$ was obtained. The little hold time plateau (highlighted by an arrow in Fig. 3) indicates the solidification of the alloy. Note that the experimental temperature of 1592 K is in reasonable agreement with the melting point of B2 predicted by the phase diagram (1583 K [56]). An overview summary of all cooling rates of the present study is given in Table 2.

**Table 2. Cooling rates (CR1, CR2 and CR3) used in the present study**

| Mould | Small mould CR1 | Y-shaped mould CR2 | Graphite crucible CR3 |
|---|---|---|---|
| Cooling rate, $K \cdot s^{-1}$ | $30 \pm 10$ | $10 \pm 5$ | $1 \pm 0.5$ |

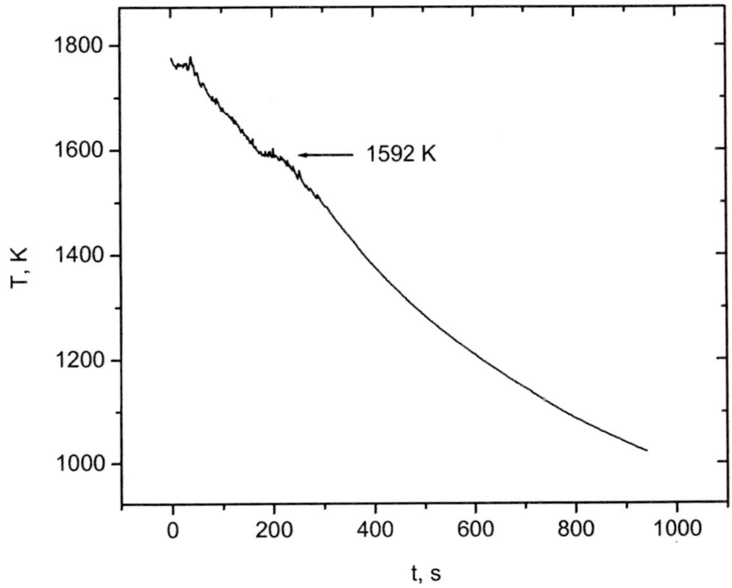


Figure 3. Experimental cooling curve for the NiTi melt kept at 1773 K for 2940 s and solidified in graphite crucible. Hold period is highlighted by an arrow.

## 3. Thermodynamic Background

Fig. 4a shows the isothermal section of the ternary Ni-Ti-C phase diagram at 1773 K. The diagram was compiled from three literature sources [54,57,58]. The liquid phase region 'L' (hatched area in Fig. 4a) extends from the Ni-corner of the diagram to the Ti-corner, where a small two phase region 'L + β-Ti' separates the L-region from the β-Ti solid phase. The melt has a limited solubility for carbon. Carbon saturated NiTi melts are in equilibrium with solid TiC ('L+TiC' two phase region). It is important to highlight that for the melts we are interested in (i.e. close to 50 at.% Ti) the elements Ti and C cannot co-exist in equilibrium. When NiTi alloys are melted in graphite crucibles there is a permanent driving force for the melts to pick-up carbon [52,53]. Fig. 4b shows the isothermal section of the Ni-Ti-C phase diagram at 1143 K [58]. There exists a two phase region "NiTi+TiC" between NiTi and TiC phases. A quasi-binary NiTi-TiC eutectic system has been presented by Du and Schuster [57] as schematically shown in Fig. 4c. Although no accurate data are available, the eutectic composition ($c_E$) is on the NiTi-side of the phase diagram. For hypoeutectic and eutectic melts, TiC forms during the eutectic reaction ("eutectic" TiC). For hypereutectic melts, the formation of primary TiC phase precedes the occurrence of the eutectic reaction.

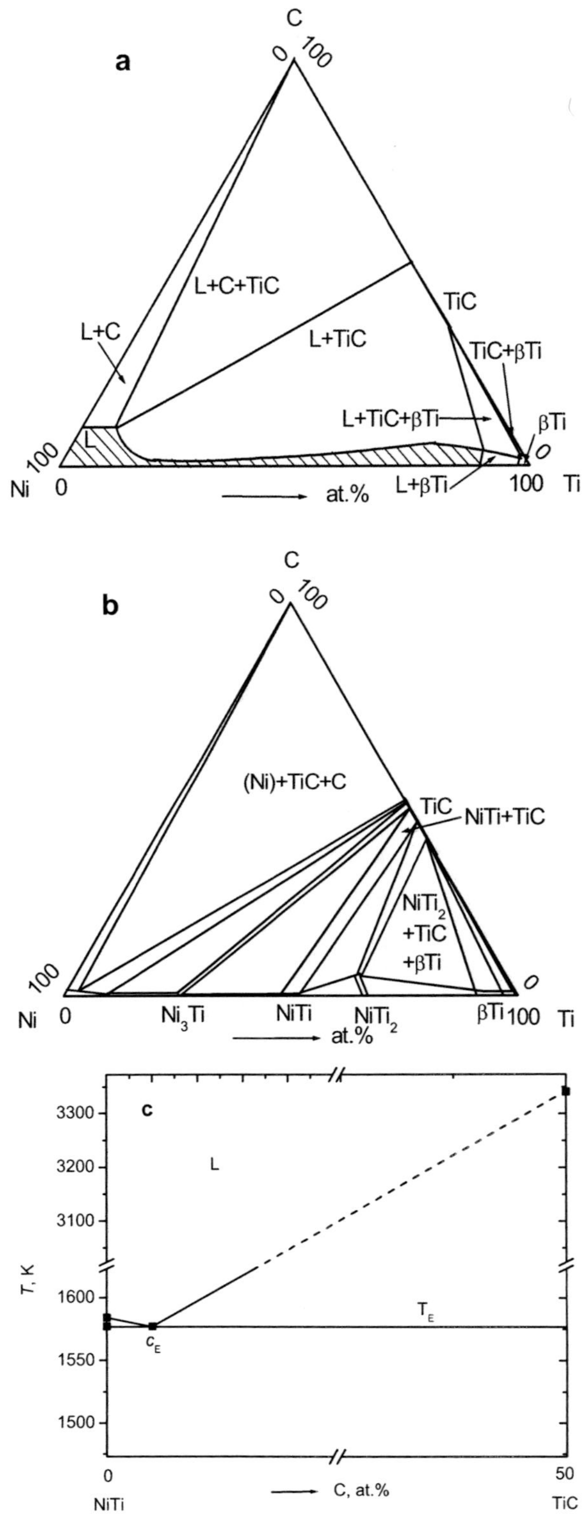


Figure 4. Ternary Ni-Ti-C phase diagram [54,57,58]. (a) Isothermal section at 1773 K, (b) isothermal section at 1143 K and (c) quasi-binary eutectic system between NiTi and TiC.

# 4. RESULTS

## 4.1. Chemical and Crystallographic Analysis

The carbon contents of the ingots investigated in this work are listed in Table 1. The chemical analysis of the ingots from the P-series shows how the carbon content of the melt increases during VIM processing with graphite crucibles, Table 1. The carbon content of the melts increases with increasing contact time between the melt and the crucible. Therefore, the kinetics of carbon enrichment in NiTi melts can be analysed based on these results, as well as the kinetics of the reaction between the melts and graphite crucibles. This has been discussed elsewhere in detail [53]. For the C-series ingots, it is difficult to accurately determine the carbon content by chemical analysis because of the inhomogeneous microstructure as will be discussed later. We simply assume that the carbon content of the three C-series experiments corresponds to the value of the last P-series ingot prior to solidification. The difference of carbon content between the two LM-series ingots results from the fact that for LM1 a high quality graphite crucible was used in combination with the cladding technique described previously [52, 53] while conventional melting in combination with a low quality graphite crucible was used for LM2.

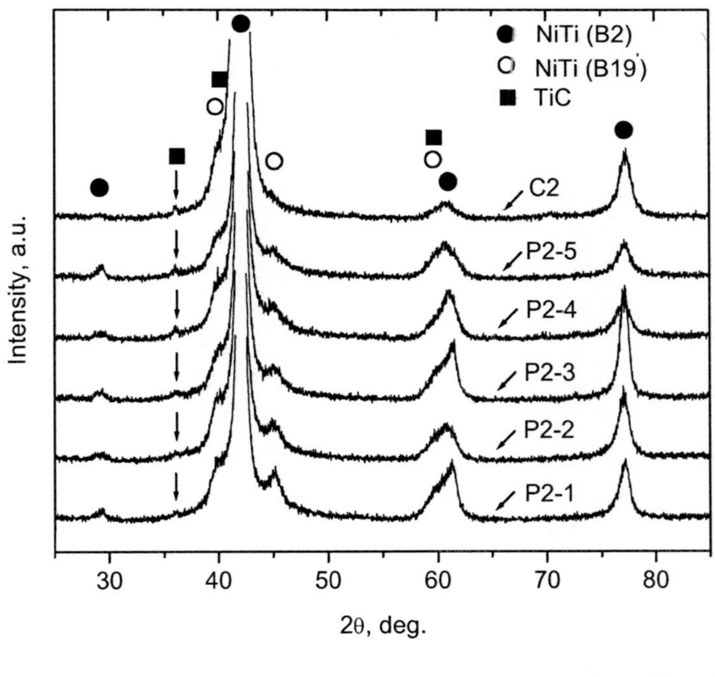


Figure 5. XRD line profiles of NiTi alloys with different amounts of carbon. The small arrows on the left of the large peaks show the presence of TiC.

According to XRD results, three phases have been identified in the P- and C-series ingots: B2 (NiTi, b.c.c. CsCl type structure, austenite), B19' (NiTi, monoclinic, martensite) and TiC. Fig. 5 shows the XRD line profiles of the NiTi-C alloys from the P2-series and C2 ingots. The diffraction peaks of the three phases overlap for two angular positions. But the little peak at about 36.1° (as highlighted by short arrows in Fig. 5) which increases in intensity with increasing carbon content of the alloy allows to distinguish TiC (111). It is obvious that B2 is the dominant phase. The presence of some B19' gives rise to the peak broadening (there is no sharp B2 (110) reflection!). While only these three phases were observed in the ingot LM1, a fourth phase, $Ni_4Ti_3$, could be detected in LM2.

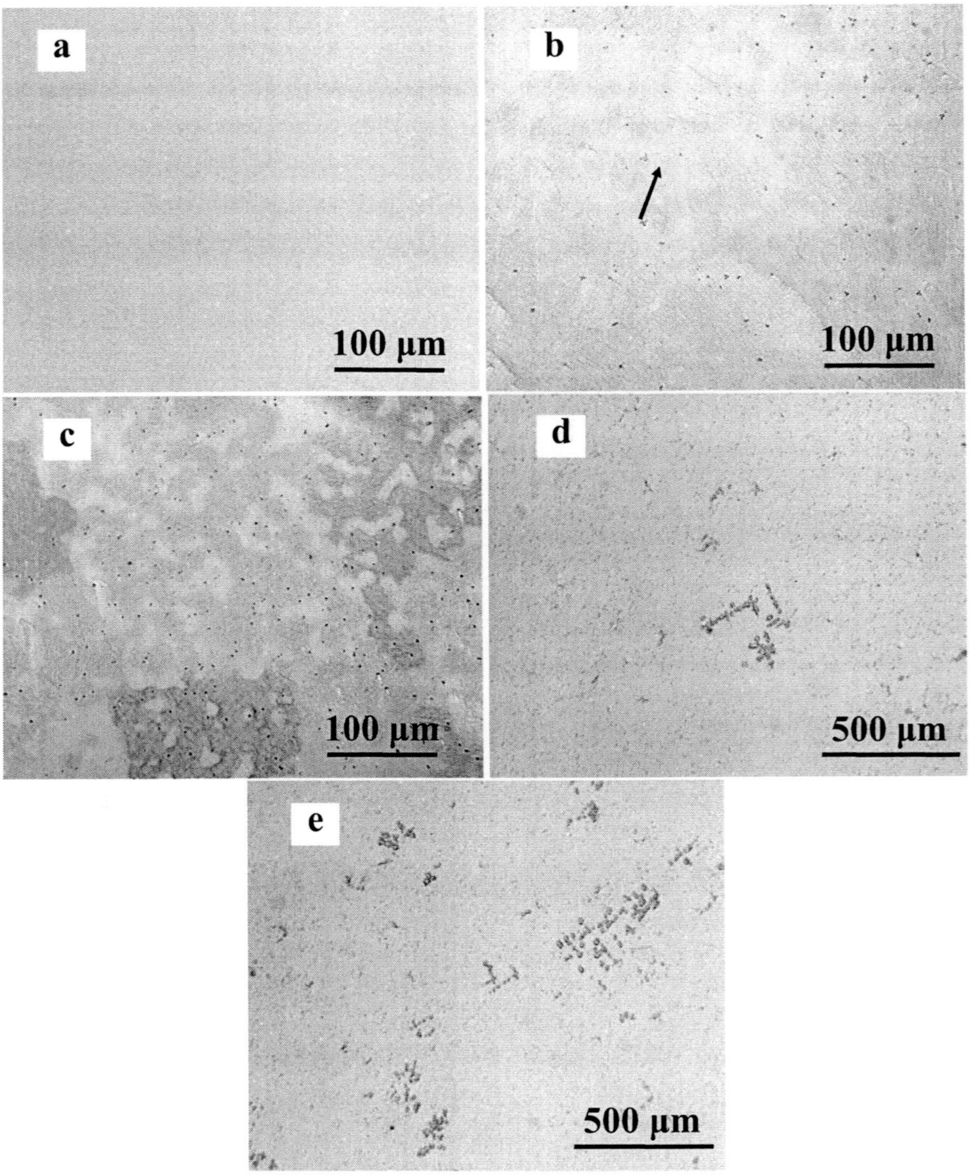


Figure 6. OM micrographs showing microstructures of NiTi alloys with different carbon contents solidified in small moulds: (a) P1-1 (1673 K, 60 s), (b) P2-2 (1773 K, 240 s), (c) P2-3 (1773 K, 540 s), (d) P3-4 (1888 K, 1500 s) and (e) P3-5 (1888 K, 2940 s).

## 4.2. Microstructures, Carbide Morphology and Crystallography

We firstly present some selected microstructures which formed in our solidification experiments, Fig. 6. Note that the carbon concentration of the NiTi alloys increases as we move from Fig. 6a to Fig. 6e. No precipitates can be detected after short reaction times at low reaction temperatures, Fig. 6a (P1-1, 1673 K, 60 s). We will later see that under the formation conditions of the ingot P1-1 small eutectic carbides form which can not be visualized using OM. When the temperature and the reaction time increase (P2-2, 1773 K, 240 s), small primary particles appear. One such a particle is highlighted by an arrow in Fig. 6b. For the ingot P2-3 (1773 K, 540 s) the amount of these primary carbides increases (Fig. 6c). In addition to small primary particles, large primary carbides (50-500 μm) have formed in the ingot P3-4 (1888 K, 1500 s), Fig. 6d. The amount of these large carbides further increases for longer contact times (P3-5, 1888 K, 2940 s), Fig. 6e. In Figs. 6d and 6e preliminary features of a dendritic carbide morphology appear.

In Fig. 7 we present two OM micrographs which were obtained from an ingot which solidified in the crucible (C3, 1888 K, 2940 s). It is important to highlight that microstructures from different locations of this ingot which cooled down at a low cooling rate look quite different. Figs. 7a and 7b were taken from specimens from the upper and lower part of the ingot, respectively. It seems reasonable to assume that the large parimary TiC dendrites in Fig. 7a have accumulated at the top of the melt due to the lower density of TiC as compared to NiTi. Moreover, this part of the ingot probably solidified last and received carbon in the course of the displacement of the solid/liquid interface during solidification. No effort was made to fully clarify this point. It is important to note that these effects are associated with slow cooling rates (C-series, cooling rates: 1 $K \cdot s^{-1}$, Table 2) and do not play a role for the small ingots (P-series, cooling rates: 30 $K \cdot s^{-1}$, Table 2).

Fig. 8 shows SEM micrographs of the microstructure of ingot P3-4 (1888 K, 1500 s). The two images shown in Fig. 8 were obtained using back scattered electrons. Fig. 8a shows a well established dendritic pattern in the center/lower right of the micrograph. The rest of the micrographs shows small primary carbides. In Fig. 8b we show the small primary carbides at a higher magnification. A very fine low contrast eutectic TiC is present in the background and can hardly be recognized.

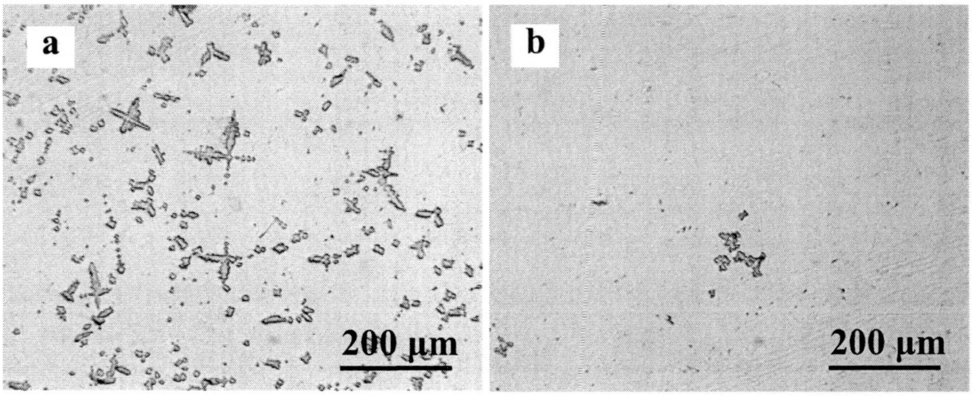


Figure 7. OM micrographs showing the microstructure of the ingot C3 (1888 K, 2940 s): (a) upper part and (b) lower part.

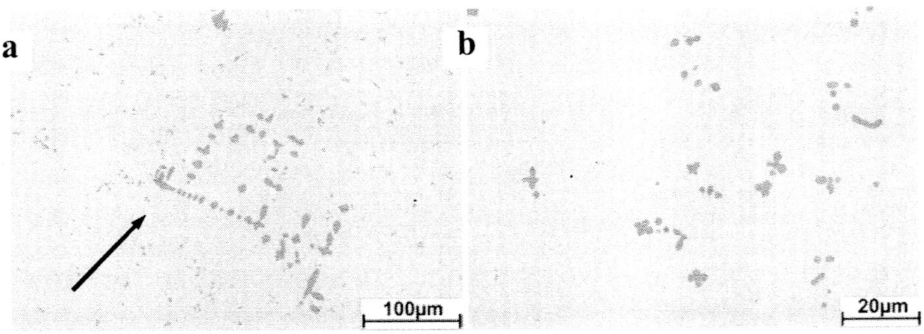


Figure 8. SEM micrographs showing the microstructure of the ingot P3-4 (1888 K, 1500 s, prepared by mechanical polishing without etching or electropolishing). (a) Large and small primary carbides and (b) small primary carbides at higher magnification.

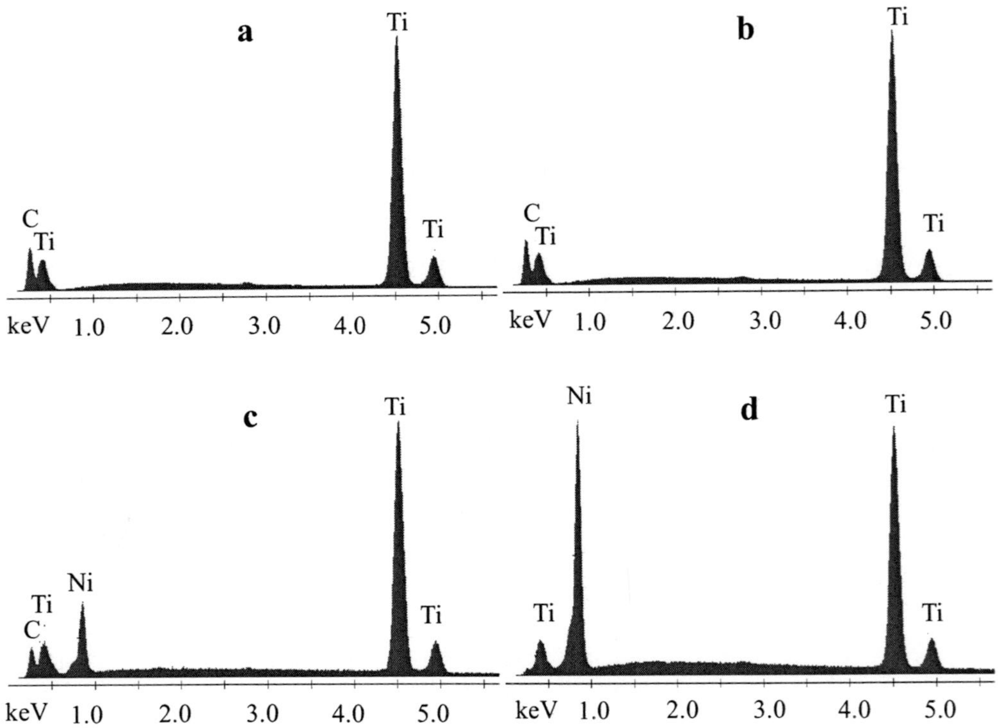


Figure 9. EDX spectra obtained from (a) a large primary carbide, (b) a small primary carbide, (c) a eutectic carbide and (d) the NiTi phase.

EDX spectra were obtained from four characteristic microstructureal regions. An EDX spectrum from a large primary TiC dendrite like the one highlighted with an arrow in Fig. 8a is shown in Fig. 9a. Note that there is no Ni-signal in the EDX spectrum of Fig. 9a. We obtain very similar EDX spectrum when analyzing the small primary TiC carbides shown in Fig. 8b, Fig. 9b. These primary particles are still larger than the lateral resolution limit of the SEM EDX method. Some carbide sizes are listed in Table 3 for reference. When we take an EDX spectrum from the fine low contrast eutectic mentioned above, we obtain the result shown in Fig. 9c. Ni is now present as an element in the spectrum, but the Ti to Ni ratio is clearly higher than 1. In contrast, the carbide free matrix shows an EDX spectrum representing a 1:1 ratio of Ni and Ti, Fig. 9d.

**Table 3. Size of small primary TiC crystals which formed under different combinations of melt temperature and cooling rate. Size given in µm**

| | 1673 K | 1773 K | 1823 K | 1888 K |
|---|---|---|---|---|
| CR1 | 1-5 | 1-10 | ---- | 1-30 |
| CR2 | ---- | ---- | 10-50 | ---- |
| CR3 | 10-15 | 30-50 | ---- | 50-100 |

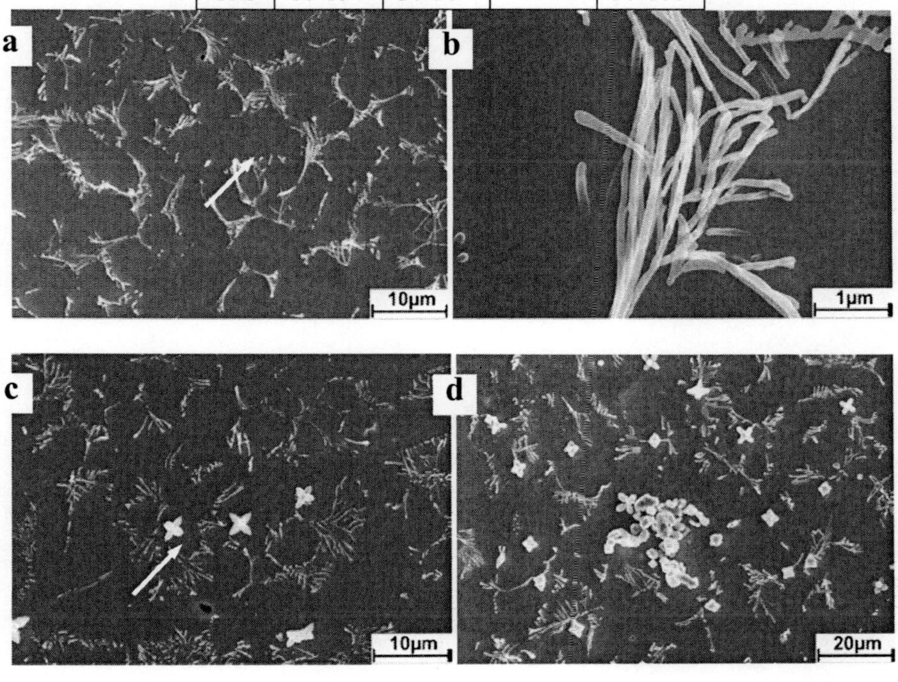


Figure 10. SEM micrographs showing microstructures of the ingots (a,b) P2-1 (1773 K, 60 s), (c) P2-2 (1773 K, 240 s) and (d) P2-4 (1773 K, 1500 s).

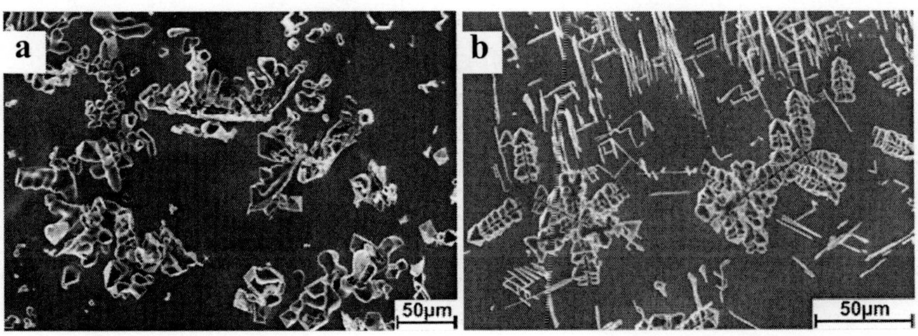


Figure 11. SEM micrographs showing microstructures of (a) upper part and (b) lower part of the ingot C3 which solidified in the graphite crucible (1888 K, 2940 s).

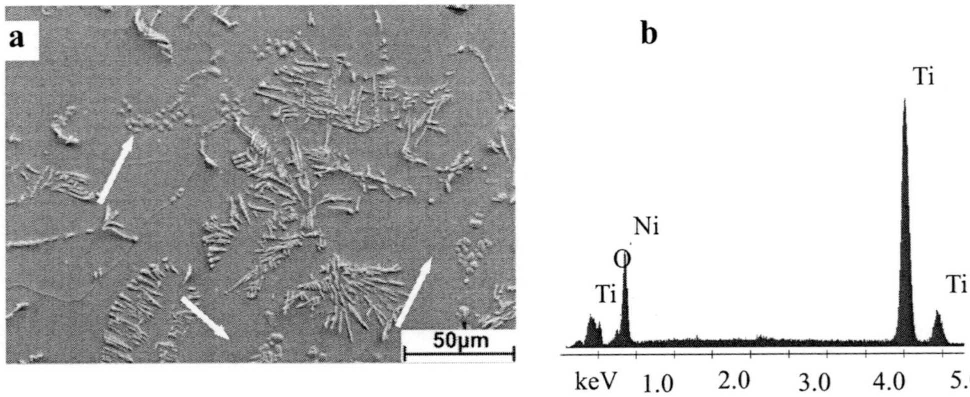


Figure 12. (a) SEM micrograph showing the microstructure of the ingot C1 (1673 K, 2940 s). Oxides are highlighted by arrows. (b) EDX spectrum from one oxide particle.

Fig. 10 shows that SEM contrast can be strongly enhanced by additional electropolishing. The four microstructures shown in Fig. 10 stem from the P2-series (1773 K). Fig. 10a shows a eutectic microstructure after a reaction time of 60 s. The area marked with an arrow in Fig. 10a is shown at a higher magnification in Fig. 10b. Fig. 10b reveals that the eutectic carbides have an elongated fibrous morphology. Fig. 10c represents a microstructure after 240 s. In addition to the eutectic microstructure some small primary carbides can be detected (one marked with an arrow). After reaction times of 1500 s microstructural regions with large primary carbides are observed (center of Fig. 10d).

Fig. 11 shows microstructures which were observed in the ingot C3 (1888 K, 2940 s, slow cooling in graphite crucible). The microstructures at the top (Fig. 11a) and at the bottom (Fig. 11b) of the ingot differ. Fig. 11a shows large primary carbides; the volume fraction of the carbides is high and the microstructure is heterogeneous. Fig. 11b shows the bottom region of the ingot where the microstructure has a lower volume fraction of carbides. The microstructure is more regular and well defined dendritic primary carbides can be observed.

Fig. 12 shows a micrograph from the ingot C1 (1673 K, 2940 s, slow cooling in graphite crucible). In this case regions with oxides can be detected. Such regions are highlighted with arrows in Fig. 12a. The corresponding EDX spectrum in Fig. 12b reveals the presence of oxygen. It seems reasonable to assume that the oxides are of type $Ti_4Ni_2O_x$ [59]. Similar microstructures were also found in the the ingots C2 (1773 K, 2940 s) and C3 (1888 K, 2940 s). In comparison, no oxide was found in microstructures of the P- and LM-series ingots, suggesting that oxygen exists in the form of solid solution in the matrix of NiTi alloys which solidified at higher cooling rates but forms the $Ti_4Ni_2O_x$ phase in the alloys which solidified at lower cooling rates.

Fig. 13 shows a microstructure from the ingot P2-5 (1773K, 2940 s). Fig. 13a shows a SEM micrograph of central primary carbides which acted as nuclei for the formation of eutectic carbide fibres. Fig. 13b provides TEM evidence for this type of process. It can be clearly seen how small fibrous eutectic carbides emerge from the cross corners of the central primary carbide (to the left and upwards). The inserted selected-area electron diffraction (SAED) pattern in Fig. 13b can be rationalized on the basis of TiC [103].

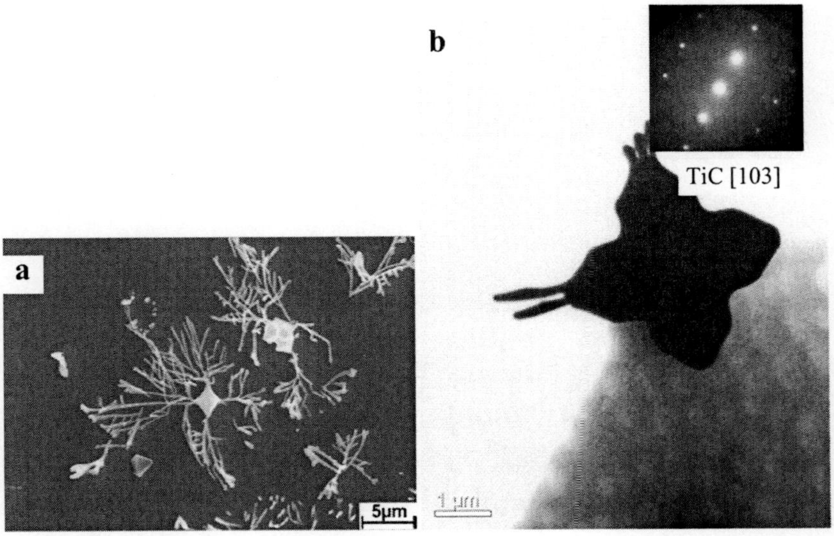


Figure 13. (a) SEM and (b) TEM bright-field (BF) micrographs showing the growth of eutectic TiC phase from primary TiC particles in the ingot P2-5 (1773 K, 2940 s). The insert in Fig. 13b shows a selected-area electron diffraction (SAED) pattern of TiC [103] zone axis from one of the eutectic carbides.

We now present a series of SEM micrographs which provide information on how small primary carbides grow, Fig. 14. Fig. 14 consists of six SEM micrographs taken from different primary carbides. But we interpret these micrographs as representing different stages of the growth process. Fig. 14a shows a primary carbide from the ingot P2-3 (1773 K, 540 s). The carbide has a size of about 2 μm and has an octahedral shape. Figs. 14b (P2-4, 1773 K, 1500 s) and 14c (P2-4, 1773 K, 1500 s) suggest that the carbide growth proceeds in all six directions defined by the six vertexes of an octahedron like the one shown in Fig. 14a. Figs. 14d (P3-1, 1888 K, 60 s), 14e (P3-2, 1888 K, 240 s) and 14f (P3-2, 1888 K, 240 s) show later stages of the growth of primary TiC phase. It can be seen that beautiful dendritic shapes eventually evolve.

We now use TEM imaging and selected-area electron diffraction to study the growth direction of carbides. Fig. 15 shows a TEM micrograph with elongated eutectic TiC. Electron diffraction was performed for three locations of the eutectic carbide. The areas where electron diffraction patterns were taken correspond to the white circles shown in the TEM micrograph of Fig. 15a. The three electron diffraction patterns are presented in Figs. 15b (Diff. 1), 15c (Diff. 2) and 15d (Diff. 3). The diffraction patterns are rotated by an angle of ~ 219° (clockwise) to account for the rotation between image and diffraction pattern for the image/diffraction conditions used. All three electron diffraction patterns show TiC crystals with the [001] zone axis close to the primary beam. For reference white arrows indicate directions of type <100> which correspond to the directions in which the elongated carbides are aligned. We conclude that all elongated fibrous parts of eutectic TiC grow in directions of type <100>. In Fig. 16 we present the same type of results for the small primary carbides shown in Figs. 16a (P3-3, 1888 K, 540 s) and 16c (P3-1, 1888 K, 60 s). The corresponding electron diffraction patterns in Figs. 16b and 16d (presented in the same way as has been explained in Fig. 15, clockwise rotation angle between image and diffraction pattern: ~ 228°) suggest that the growth of small primary carbides also follows <100> directions.

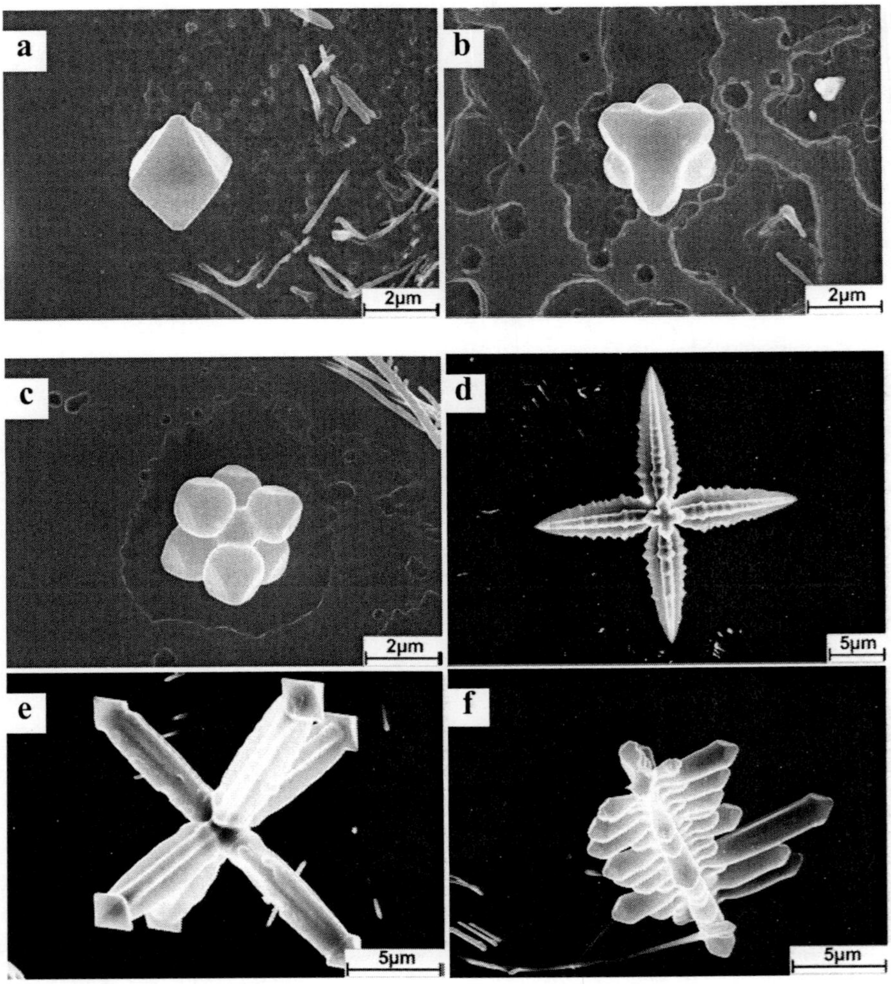


Figure 14. Series of SEM micrographs showing transition from octahedral to dendritic carbides. (a) P2-3 (1773 K, 540 s), (b,c) P2-4 (1773 K, 1500 s), (d) P3-1 (1888 K, 60 s) and (e,f) P3-2 (1888 K, 240 s).

In the present report we only show some selected micrographs. But we have investigated all material states listed in Table 1. The microstructural information on the material states which were not explicitly presented is briefly summarized in the last column of Table 1, where "NiTi" refers to the matrix of the NiTi SMAs, "E" indicates the presence of a eutectic TiC phase and "SP" and "LP" indicate the presence of small and large primary carbides, respectively. Moreover, "OX" indicates the detection of oxides of type $Ti_4Ni_2O_x$ in the microstructure. In Tables 3 and 4 we made an attempt to briefly summarize all our experimental findings. Table 3 shows how characteristic carbide sizes depend on cooling rates (CR1, CR2 and CR3 specified in Table 2) and the initial temperature of the melt. Table 4 makes an attempt to qualitatively relate the volume fraction of different types of small primary carbides (octahedral and dendritic) present in the microstructure. The first row of Table 4 has three columns. The three "+" signs in the first column indicate that about 75% of all carbides are in an early state of growth ("octahedral", like the one shown in Fig. 14a). The single "+" sign in the second column indicates that about 25% of all carbides have grown into dendritic shapes ("dendritic", like shown in Figs. 14d, 14e and 14f). And the four "+" signs in the third column provide the information that the carbide density is high (very low: +, very high: ++++++). Thus the second row indicates that only primary carbides in an early growth state were detected and that the overall carbide density is very low. The other rows of Table 4 provide similar information.

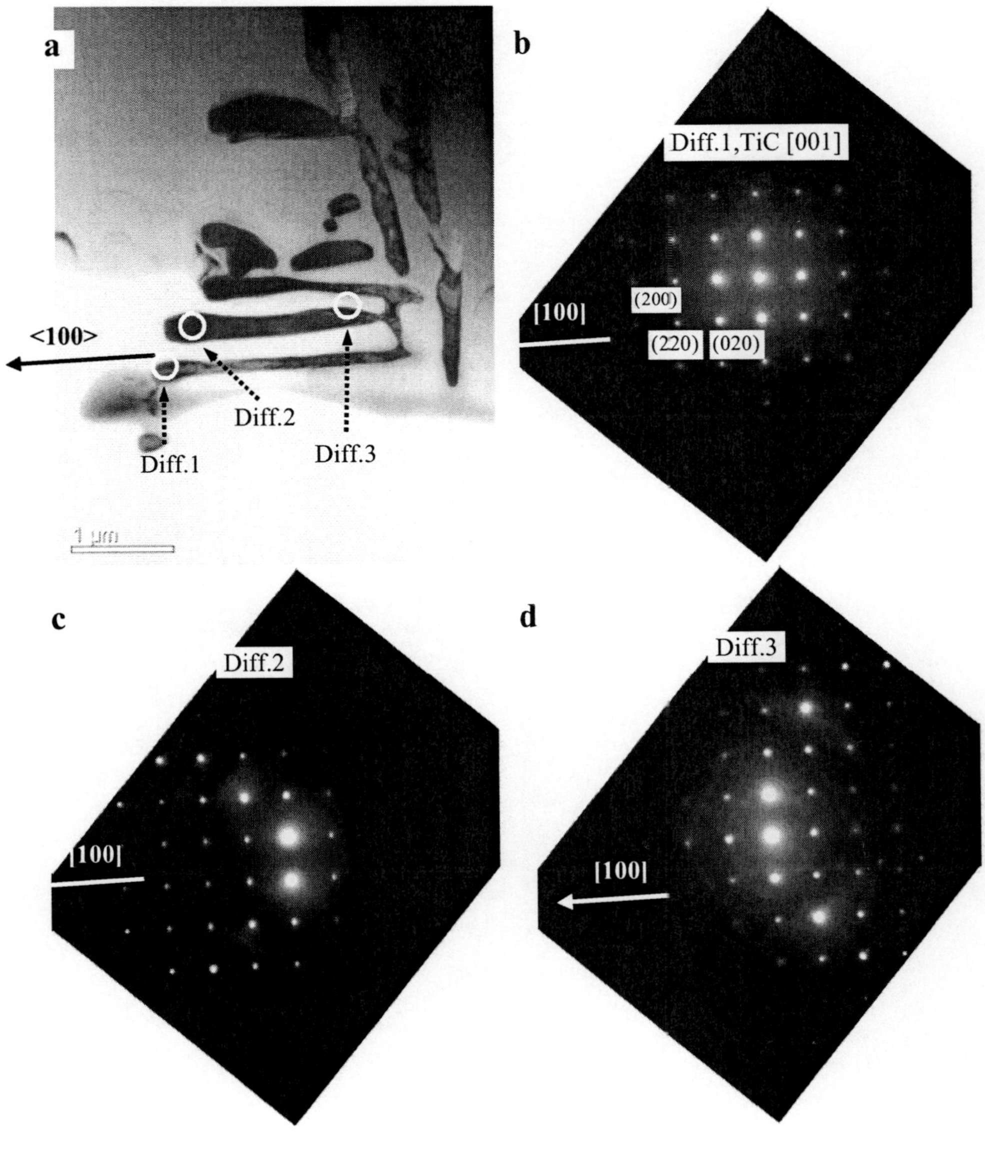


Figure 15. (a) TEM micrograph showing eutectic TiC phase in the ingot P2-5 (1773 K, 2940 s). (b,c,d) selected area electron diffraction patterns of the three locations highlighted in Fig. 15a.

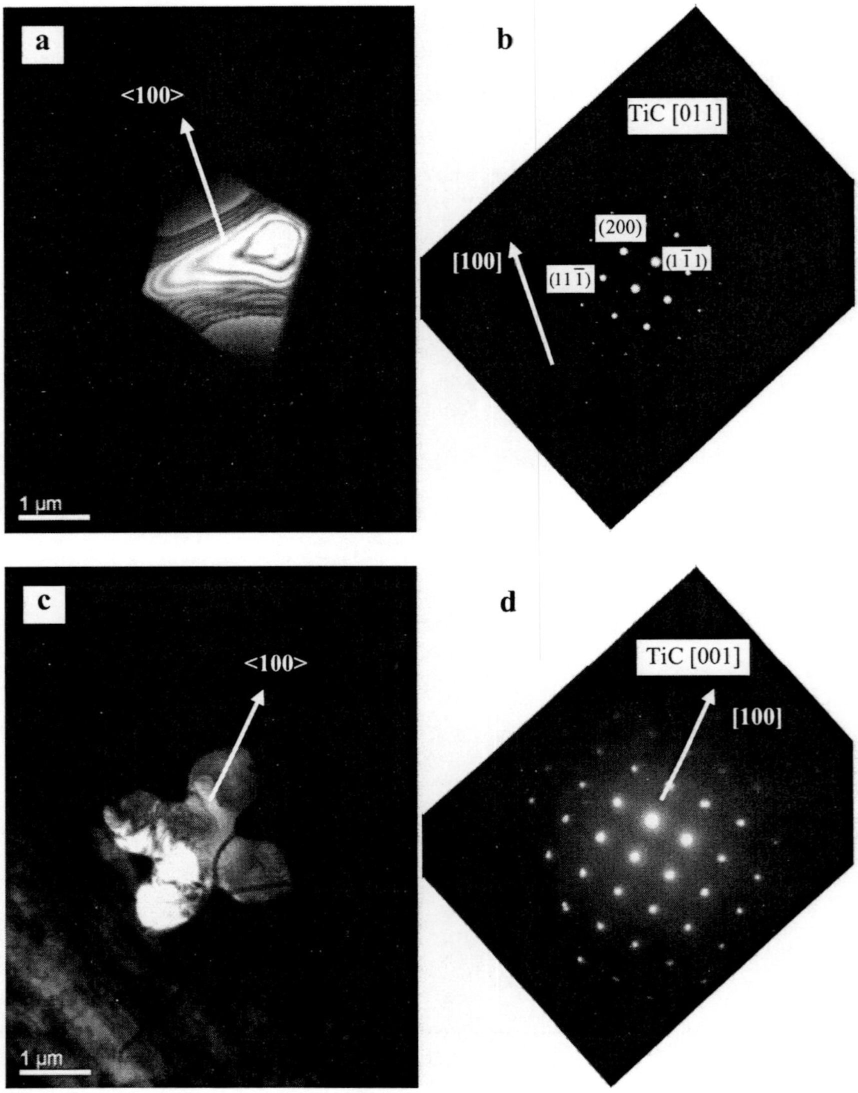


Figure 16. (a,c) TEM micrographs showing small primary TiC particles in the ingots P3-3 (1888 K, 540 s) and P3-1 (1888 K, 60 s), respectively; (b,d) corresponding selected area electron diffraction patterns.

**Table 4. Overview summary of relative and absolute quantities of different types of small primary carbides. (+: very few, ++++++: a lot, others: in between)**

| | | octahedral | dendritic | total |
|---|---|---|---|---|
| 1673 K | CR1 | +++ | + | ++++ |
| | CR3 | + | | + |
| 1773 K | CR1 | ++ | ++ | ++++ |
| | CR3 | | + | + |
| 1888 K | CR1 | + | +++++ | ++++++ |
| | CR3 | | + | + |
| 1823 K | CR2 | + | + | ++ |

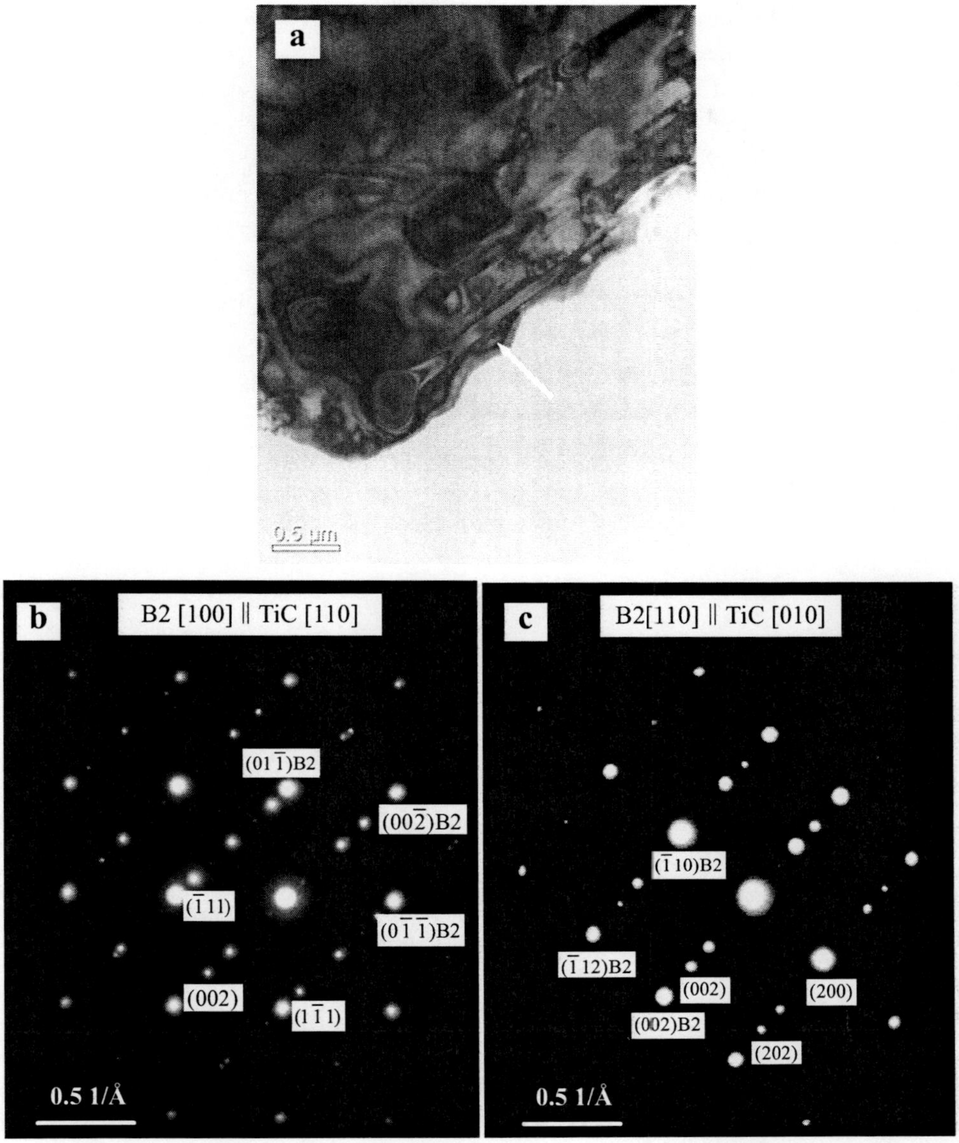


Figure 17. (a) TEM micrograph showing a eutectic TiC fibre in the ingot LM1; (b,c) corresponding SAED patterns.

## 4.3. Orientation Relationships between TiC And B2 Matrix

We now study the orientation relationship between the elongated eutectic carbide as highlighted by an arrow in Fig. 17a (LM1, 1823 K, 10 s) and the surrounding B2 matrix. The electron diffraction patterns presented in Figs. 17b and 17c were obtained at two distinct tilt positions. Fig. 17b shows a tilt position where the [100] direction of the B2 phase is parallel to the [110] direction of TiC. Fig. 17c shows the diffraction data of both phases when the specimen is tilted by 45° into a position where the [110] direction of the B2 phase is parallel to the [010] direction of TiC. The results shown in Fig. 17 are crystallographically consistent. However, the results shown in Fig. 18 also need to be considered. The microstructure shown in Fig. 18a represents the ingot LM2 (1823 K, 10 s). Fig. 18a shows two particles (which probably are part of one eutectic carbide, because they have exactly the same diffraction patterns). In Fig. 18b and 18c diffraction patterns were taken from areas which include matrix and carbide phases. The diffraction pattern presented in Fig. 18b shows the same results which were obtained before. The [100] direction of the B2 phase is parallel to the [110] direction of TiC. In Fig. 18c the specimen was tilted such that the primary beam is parallel to the [011] direction of the B2 phase. On a first glance diffraction spots from TiC appear to be absent. But when a small selected area diffraction is used to focus on TiC we can identify two spots which are clearly associated with TiC (220) and TiC $(\overline{2}\overline{2}0)$, Fig. 18d. In addition, examinations on many TEM specimens show that only some of the elongated, fibrous eutectic carbides exhibit a specific orientation relationship with the B2 matrix. Furthermore, no relationship was found between primary TiC phase and the B2 matrix.

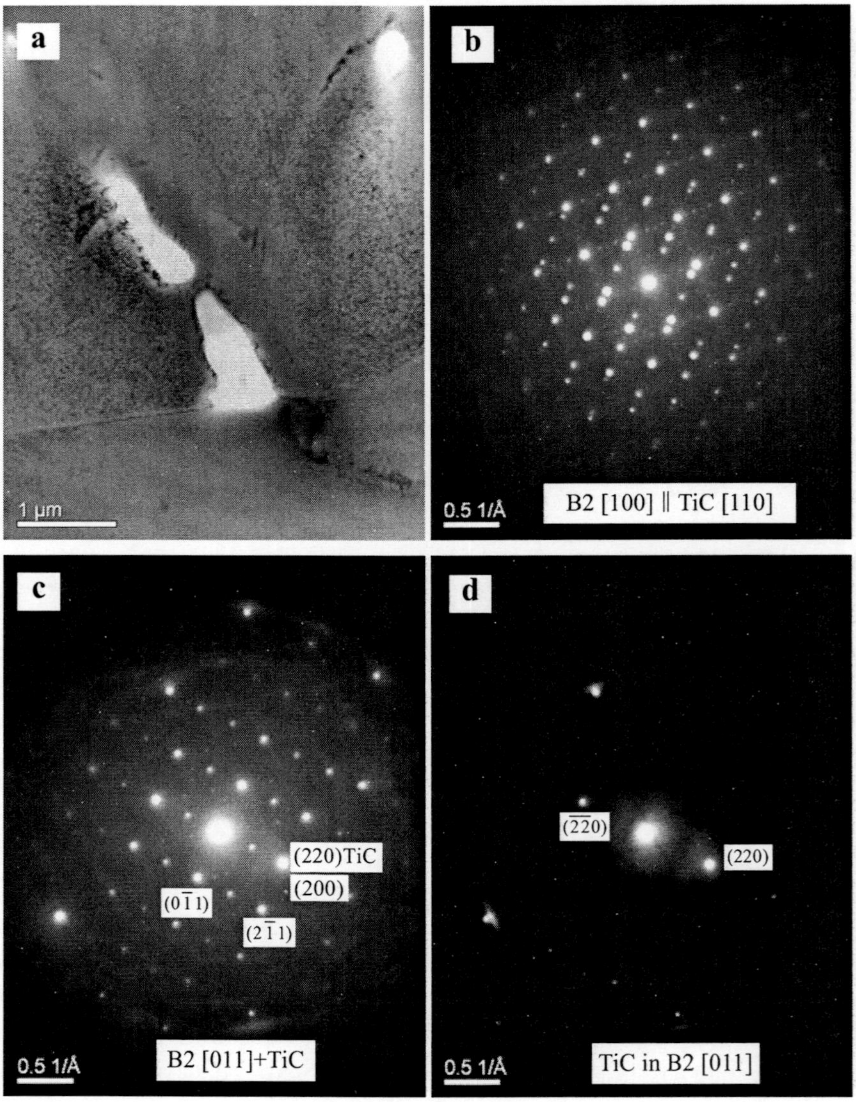


Figure 18. TEM micrograph showing a eutectic TiC fibre in the ingot LM2; (b,c,d) corresponding SAED patterns.

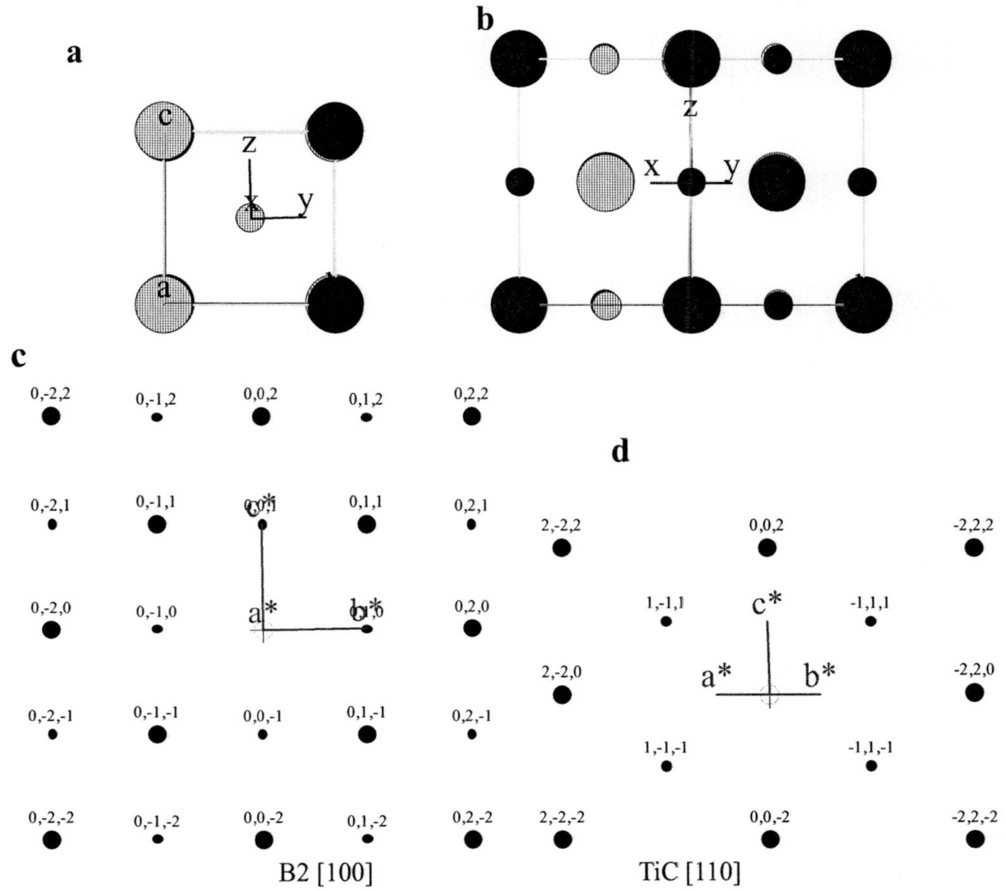


Figure 19. Schematic of (a,b) lattices and (c,d) SAED patterns corresponding to the orientation relationship: (a,c) B2 [100] || (b,d) TiC [110]. ( In (a), big circles: Ni or Ti atoms; small circle: Ti or Ni atom; in (b) big circles: Ti or C atoms; small circles: C or Ti atoms; in both, hatched circles: atoms in parallel planes of B2 (110) and TiC (020).).

We now use a crystallographic software (CaRIne v3.1) to visualize the position of atoms (in real space) and the position of diffraction spots (in reciprocal space). Fig. 19 rationalizes the experimental results shown in Figs. 17b and 18b. The crystal lattices of B2 and TiC are shown in Figs. 19a and 19b, where the [100] direction of B2 is parallel to the [110] direction of TiC. The positions of the corresponding electron diffraction spots are shown in Fig. 19c for B2 [100] and Fig. 19d for TiC [110]. It is clear that a superposition of Figs. 19c and 19d fully rationalizes the diffraction patterns shown in Figs. 17b and 18b. Fig. 20 provides the same type of crystallographic information (position of real and reciprocal lattice points) for the experimental results presented in Fig. 17c. Figs. 20a and 20b present the orientations of the B2 matrix ([110]) and TiC ([010]). And the superposition of the two dimensional cuts through the corresponding reciprocal lattices (Figs. 20c and 20d) completely rationalizes the positions of all diffraction spots in Fig. 17c. The orientation relationship between the B2 matrix and the TiC crystal is such that B2 [100] is parallel to TiC [110] and B2 [110] is parallel to TiC [010]. Note that these two lattice relations fix the positions of the atoms of both phases in space with respect to each other. Fig. 21 represents crystallographic information which helps to interpret the experimental results shown in Fig. 18c. When we orient the B2 matrix in a [011] direction (Fig. 21a) then the corresponding projection of TiC no longer yields a simple arrangement of atoms (like in the case of Figs. 19 and 20). In this orientation the TiC lattice only has one type of well defined closely spaced lattice planes of type (220) which account for the two diffraction spots close to the primary beam in Fig. 18d. These two spots overlap with the spots of the B2 reciprocal lattice in Fig. 18c and this rationalizes the experimental findings discussed earlier.

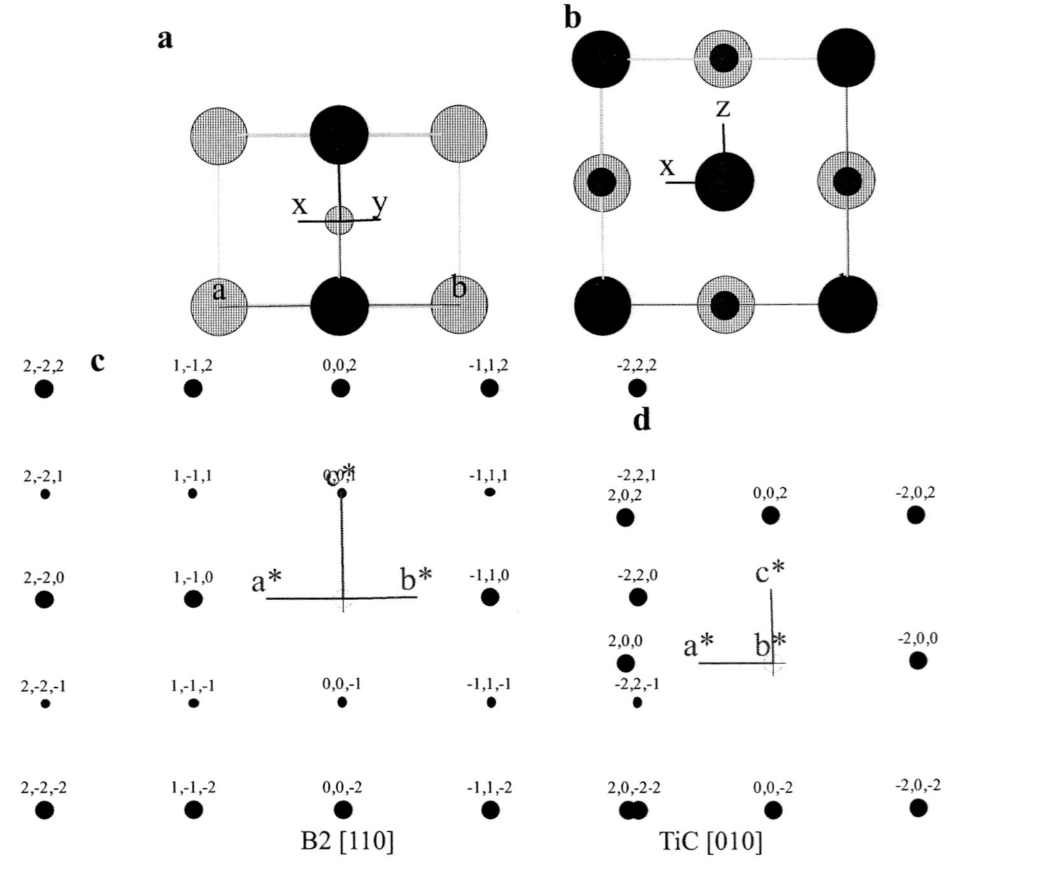


Figure 20. Schematic of (a,b) lattices and (c,d) SAED patterns corresponding to the B2 [110] pole: (a,c) B2 [110]||(b,d) TiC [010]. (Symbols: the same as those in Fig. 19.)

## 5. DISCUSSION

The metallographic results obtained in the present investigation show that TiC crystals always form when NiTi SMAs are produced using VIM in graphite crucibles. However, the amount, size, morphology and distribution of carbides strongly depend on the carbon content of the melt, the temperature of the melt prior to solidification and cooling rate. Moreover other parameters like stirring of the melt and associated detachment processes from a thin TiC layer at the graphite/melt interface need to be considered [52, 53]. Larger agglomerates of carbides (large primary TiC as shown in Figs. 6d, 6e, 8a, 10d and 11a) can be a result of a direct growth on a pre-existing TiC particle resulting from a detachment process (from the TiC layer on the graphite crucible). The present investigation documents different types of carbide microstructures which characterize NiTi alloys after vacuum induction melting in graphite crucibles. Our results have also shown that there are cases where ingots can have strongly heterogeneous microstructures. This can for example be a consequence of the difference in densities between carbides and the melt.

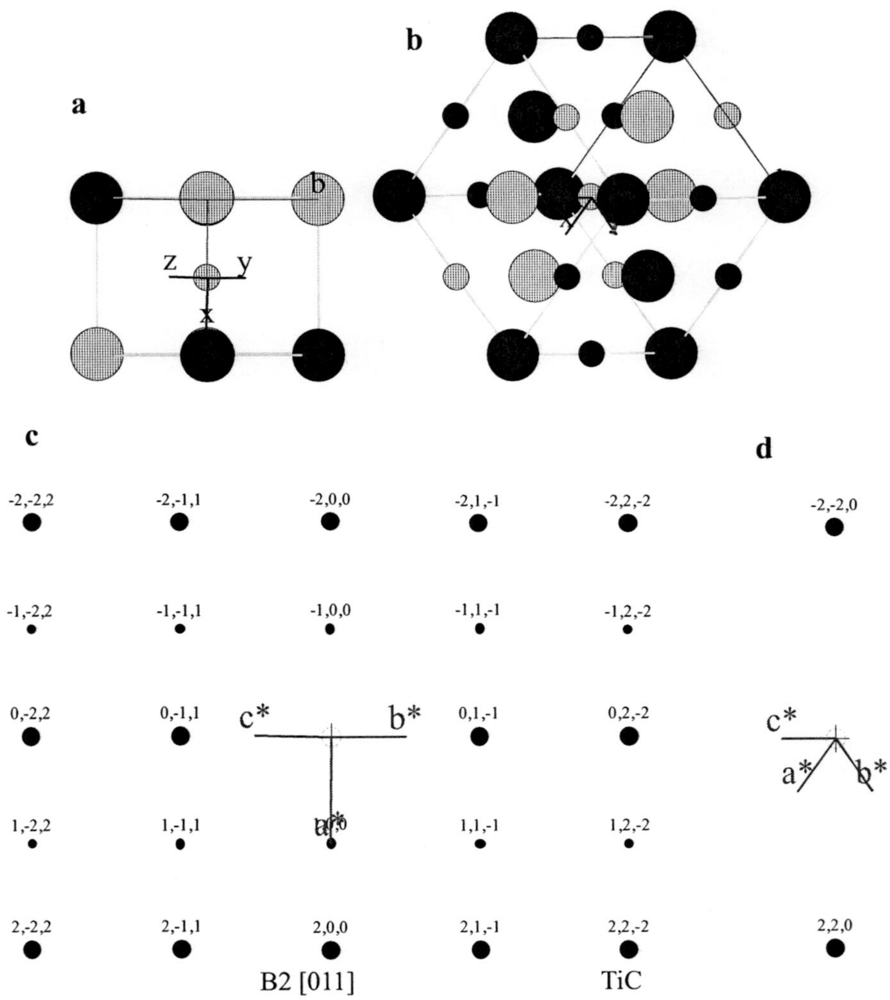


Figure 21. Schematic of (a,b) lattices and (c,d) SAED patterns corresponding to the B2 [011] pole. (a,c) B2 [011]; (b,d) TiC. (Symbols: the same as those in Fig. 19.).

In our study we have identified three categories of TiC in addition to large primary carbides. There are carbides which seem to form as a result of a eutectic reaction on the basis of the quasi-binary NiTi-TiC eutectic system [57]. Such carbides are shown in Figs.10, 11b, 12a, 13a, 15a, 17a and 18a. These carbides have elongated, fibrous shapes and are too small to be detected under the present conditions (without etching or electropolishing) using optical microscopy. The direction of the elongated carbides was shown to be <100>. In this respect TiC behaves like many other face centered cubic (fcc) crystals: the <100> direction represents a preferred crystal growth direction [60]. The fact that parallel bundles of eutectic carbides are observed (see for example Fig. 11b) suggests that the eutectic reaction can proceed in a way which has been described for the formation of pearlite [61]. This process is characterized by a simultaneous growth of both phases and a small scale diffusion of carbon, in our case from oversaturated undercooled NiTi to TiC. There also seem to be other eutectic regions where the melt simply decomposes into the two phases. Our microstructural evidence does not allow to decide on the precise nature of the eutectic reaction. To our knowledge we are the first to describe the presence of this type of eutectic microstructure in NiTi SMAs. Some (not all) of the eutectic carbides have a specific orientation relation to the matrix which is given by the [100] and [110] directions of B2 being parallel to the [110] and [010] directions of TiC.

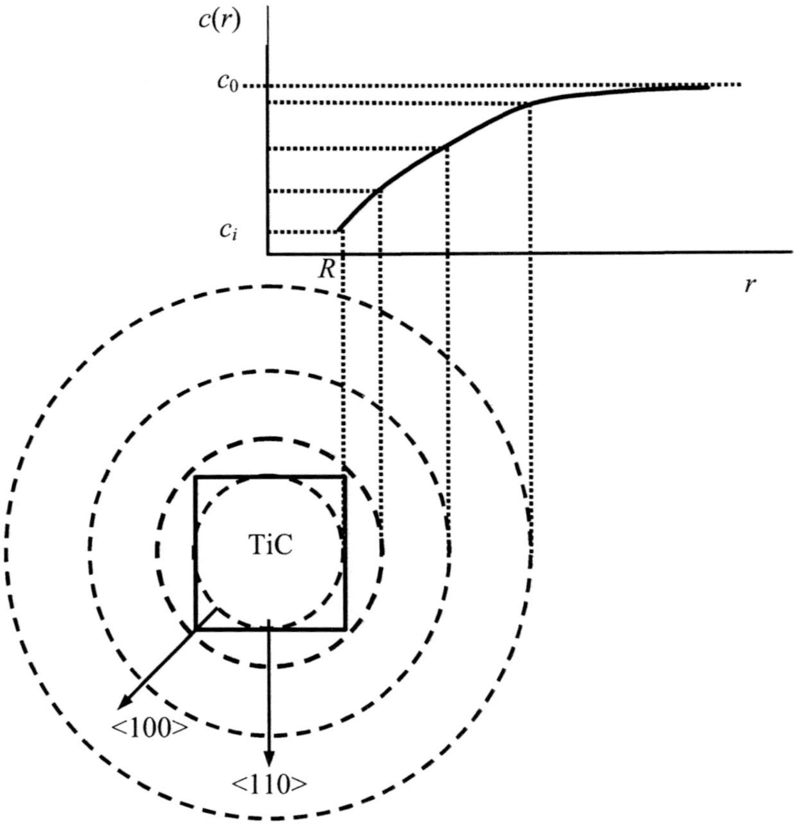


Figure 22. Carbon concentration in front of a growing carbide and varying thermodynamic conditions when the carbide shape is not spherical (modified from Ref. [12]).

The other two types of carbides which we have observed in the present study represent small primary carbides (which form in a hypereutectic melt when cooling through a two phase region) in earlier and later stages of growth. Examples of primary carbides in the early stages of growth are given in Figs. 10d, 13a, 14a, 14b and 16a. These carbides have an octahedral shape and are therefore referred to as octahedral carbides in the present work.

Examples of primary carbides in the later stages of growth are presented in Figs. 10c, 10d, 11b, 13a, 13b, 14c-14f and 16c. These carbides have beautiful dendritic shapes and are referred to as dendritic carbides.

Small octahedral carbides like the one shown in Fig 14a have interfaces which most probably represent {111} planes and <100> directions present preferred growth directions. In an early stage of growth this octahedral shape is maintained. When we look carefully at our results we find that octahedral particles can only grow to a certain size which depends on initial melt temperature and cooling rate (see Tables 3 and 4). It is difficult to draw one clean set of sharp conclusions. But it seems reasonable to assume that in the early stages of growth of primary TiC the system attempts to establish TiC {111} surfaces which are in contact with the melt. This may have to do with a low value of the interfacial energy for this type of interface. Moderate growth in <100> directions allows to maintain TiC {111} surfaces in contact with the melt. In this explanation we combine an energetic (interfacial energy) and kinetic argument (easy growth direction). We suggest that larger dendritic carbides like those shown in Fig. 14d-14f, start growing from smaller octahedral precursors as suggested by the series of SEM micrographs presented in Fig. 14. We suggest that additional geometric/kinetic arguments must be considered to discuss the transition from octahedral to dendritic growth. It is important to highlight that the carbon concentration in the environment of a growing carbide decreases and further growth is limited by the diffusion of carbon from the melt to the carbide/melt interface. Fig. 22 schematically illustrates (i) that there is a carbon concentration profile in front of the TiC/melt interface and (ii) that in case of non-spherical particle sharp edges reach out to melt locations with higher carbon concentrations. This eventually results in an instability of a planar solid/liquid interface and is associated with constitutional undercooling [60]. It seems reasonable to assume that a mechanism like the one proposed in Fig. 23 governs the evolution of dendritic carbides.

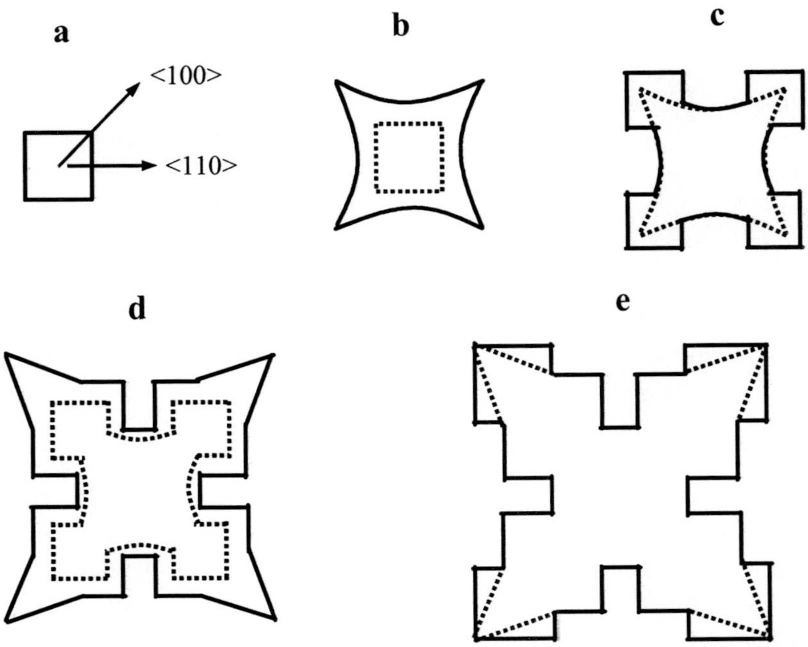


Figure 23. Schematic illustration of the evolution of a dendritic carbide [12, 60].

The present work is the first systematic report on the formation of TiC crystals in carbon-containing NiTi shape memory alloy melts although the formation of TiC in other melts has been reported in the literature [3-6, 20-22]. Several atomistic and morphological aspects of the formation of TiC were reported [3-6, 12-14]. Jin et al. [12,14] explained the formation and growth of TiC crystals in $TiC_p$/Ti composites using coordination polyhedron growth unit theory which was proposed by Zhong et al. [62]. These previous results could not be directly used in the present study because key energetic and kinetic parameters change from one system to another.

## 6. Conclusion

In the present study we investigate the formation of TiC during vacuum induction melting of NiTi shape memory alloys in graphite crucibles. We performed melt experiments at temperatures ranging from 1673 to 1888 K with durations between 10 and 2940 s and different cooling rates. The resulting ingots were investigated with respect to their microstructure (using optical, scanning and transmission electron microscopy) and with respect to the phases and characteristic crystallographic directions (using X-ray and electron diffraction). The following results were obtained.

(1) During graphite crucible vacuum induction melting of NiTi shape memory alloys, TiC crystals always form. The volume fraction of carbides, their distribution in the ingot, their size and morphology strongly depend on parameters like melt temperature, duration of the process, carbon concentration of the melt and cooling rate during solidification. Very different microstructures can be obtained.

(2) The NiTi-C system has a eutectic reaction where a NiTi/TiC eutectic microstructure forms. This microstructure is present in all ingots. However, under the conditions used in the present study it is too fine to be resolved using optical microscopy. It can be easily investigated using scanning and transmission electron microscopy. Eutectic TiC can be elongated and fibrous. The fibre direction was found to be TiC <100>. Some of the eutectic carbides show a well defined lattice relationship with respect to the NiTi matrix. Other types of eutectic microstructures were also observed.

(3) For NiTi alloys with higher carbon concentrations small and large primary carbides formed during solidification. Two types of small primary TiC phase were observed, smaller octahedral and larger dendritic carbides. The preferred growth direction of these octahedral and dendritic TiC crystals was found to be <100>.

(4) The formation and growth of octahedral TiC crystals was discussed on the basis of energetic (low interfacial energies for TiC {111} and kinetic arguments (TiC <100> represents an easy growth direction). The size to which the octahedral carbides can grow is limited and depends on the initial melt temperature and cooling rate.

(5) There is a moment where further growth of octahedral carbides occurs such that beautiful dendritic TiC crystals form. To discuss the morphology of these carbides the thermodynamic conditions in the melt surrounding the carbide (carbon concentration and constitutional undercooling) also need to be considered.

## ACKNOWLEDGEMENT

Z.H. Zhang acknowledges funding by the Alexander von Humboldt Foundation (AvH). All authors acknowledge funding from the Deutsche Forschungsgemeinschaft (DFG) through project C7 of SFB 459. Support by the Land NRW is also gratefully acknowledged.

## REFERENCES

[1] Lengauer, W. In *Handbook of Ceramic Hard Materials*; Riedel, R.; Ed.; Transition metal carbides, nitrides and carbonitrides; Wiley-VCH: Weinheim, DE, 2000; Vol. 1, pp 202-252.
[2] Chien, F.; Nutt, S.; Cummings, D. *Phil. Mag. A* 1993, 68, 325.
[3] Strzeciwilk, D.; Wokulski, Z. *Cryst. Res. Technol.* 1999, 34, 777-784.
[4] Strzeciwilk, D.; Tkacz, P.; Wokulski, Z. *Cryst. Res. Technol.* 2000, 35, 1295-1303.
[5] Strzeciwilk, D.; Wokulski, Z.; Tkacz, P. *J. Alloys Comp.* 2003, 350, 256-263.
[6] Strzeciwilk, D.; Wokulski, Z.; Tkacz, P. *Cryst. Res. Technol.* 2003, 38, 283-287.
[7] Qi, S.R.; Huang, X.T.; Gan, Z.W.; Ding, X.X.; Cheng, Y. *J. Cryst. Growth* 2000, 219, 485-488.
[8] Bai, Y.J.; Feng, X.; Lü, B.; Wang, C.G.; Qi, Y.X.; Liu, Y.X.; Zhu, B.; Wang, Y.X. *Chem. Phys. Lett.* 2004, 388, 58-61.
[9] Williams, W.S. *JOM* 1997, 49, 38.
[10] Strzeciwilk, D.; Wokulski, Z. *Arch. Nauk. Mater.* 1999, 2, 123.
[11] Jarfors, A.; Fredrikson, H. *Microgravity Sci. Technol.* 1991, III/4, 216.
[12] Jin, Y.X.; Wang, H.W.; Zeng, S.Y.; Zhang, E.L. *Trans. Nonferrous Met. Soc. China* 2002, 12, 1158-1163.
[13] Jin, Y.X.; Zeng, S.Y.; Zhang, E.L.; Zhu, Z.J. *Rare Metal Mater. Eng.* 2003, 32, 451-455.
[14] Jin, Y.X.; Zhang, H.; Zeng, S.Y.; Zhang, E.L.; Li, Q.F. *Acta Metall. Sin.* 2002, 38, 1223-1227.
[15] Otani, S.; Ishizawa, Y. *Prog. Cryst. Growth Charact.* 1992, 23, 153.
[16] Kato, A.; Tamari, N. *J. Cryst. Growth* 1980, 49, 199.
[17] Wokulski, Z.; Wokulska, K. *J. Cryst. Growth* 1983, 62, 439.
[18] Christensen, A. *J. Cryst. Growth* 1976, 33, 99.
[19] Otani, S.; Honma, S.; Tanaka, T.; Ishizawa, Y. *J. Cryst. Growth* 1983, 61, 1.
[20] Lundström, T.; Korsukova, M.M.; Gurin, V.N. *Prog. Cryst. Growth Charact.* 1988, 16, 143.
[21] Shimada, S.; Watanabe, J.; Kodaira, K.; Matsushita, T. *J. Mater. Sci.* 1989, 24, 2513.
[22] Higashi, I.; Takahashi, Y.; Atoda, T. *J. Cryst. Growth* 1976, 33, 207.
[23] Saburi, T. In *Shape Memory Materials*; Otsuka, K.; Wayman, C.M.; Eds.; Cambridge University Press: Cambridge, UK, 1998; pp 49-73.
[24] Otsuka, K.; Ren, X. *Intermetallics* 1999, 7, 511-528.
[25] Van Humbeeck, J. *Mater. Sci. Eng. A* 1999, 273-275, 134-148.
[26] Duerig, T.; Pelton, A.; Stöckel, D. *Mater. Sci. Eng. A* 1999, 273-275, 149-160.
[27] Otsuka, K.; Kakeshita, T. *MRS Bull.* 2002, 27, 91-100.

[28] Tang, W.; Sundmann, B.; Sandström, R.; Qiu, C. *Acta Mater.* 1999, 47, 3457.
[29] Miyazaki, S. In *Engineering Applications of Shape Memory Alloys*; Duerig, T.W.; Melton, K.N.; Stöckel, D.; Wayman, C.M.; Eds.; Butterworth-Heinemann: London, UK, 1990; p 394.
[30] Morgan, N.B.; Friend, C.M. *J. Phys. IV France* 2001, 11, Pr8-325.
[31] Duerig, T.W. *MRS Bull.* 2002, 27, 101-104.
[32] Sawaguchi, T.; Kausträter, G.; Yawny, A.; Wagner, M.; Eggeler, G. *Metall. Mater. Trans. A* 2003, 34, 2847-2859.
[33] Eggeler, G.; Hornbogen, E.; Yawny, A.; Heckmann, A.; Wagner, M. *Mater. Sci. Eng. A* 2004, 378, 24-33.
[34] Benard, W.L.; Kahn, H.; Heuer, A.H.; Huff, M.A. *J. MEMS* 1998, 7, 245–51.
[35] Takeuchi, S.; Shimoyama, I. *J. MEMS* 2000, 9, 24–31.
[36] Shabalovskaya, S.A. *Bio-Med. Mater. Eng.* 1996, 6, 267.
[37] Ayers, R.A.; Simske, S.J.; Bateman, T.A.; Petkus, A.; Sachdeva, R.L.C.; Gyunter, V.E. *J Bio-Med. Mater. Res.* 1999, 45, 42–47.
[38] Firstov, G.S.; Vitchev1, R.G.; Kumar, H.; Blanpain, B.; Van Humbeeck, J. *Biomaterials* 2002, 23, 4863.
[39] Zhu, J.H.; Liu, C.T. *Scripta Mater.* 1999, 41, 55.
[40] Drennen, D.C.; Jackson, C.M.; Wagner, H.J. *The Development of Melting and Casting Procedures for Nitinol Nickel-Base Alloys*; Research Report Sandia Laboratories, SC-CR-69-3070: Albuquerque,US, 1968.
[41] Jackson, C.M.; Wagner, H.J.; Wasilewski, R.J. *55-Nitinol - The Alloy with a Memory: Its Physical Metallurgy, Properties, and Application*; NASA-SP 5110: Washington DC, US, 1972.
[42] Butalov, L.V.; Gajdaj, P.I.; Zakharova, N.N.; Kruglov, B.V.; Filin, Y.A. *Litejnoe Proizvod*, 1980, 7, 11 (in Russian).
[43] Honma, T. In *Shape Memory Alloys*; Funakubo, H.; Ed.; Gordon and Breach: New York, US, 1987; pp 89-101.
[44] Stöckel, D.; Hornbogen, E.; Ritter, F.; Tautzenberger, P. *Legierungen mit Formgedächtnis*; Expert Verlag: Ehningen bei Böblingen, DE, 1988; p 45 (in German).
[45] Ehrenstein, H. *Die Herstellung und das Formerinnerungsvermögen von NiTi*; Dr.-Ing. Thesis; TU Berlin: Berlin, DE, 1985 (in German).
[46] Russel, S.M. In *Proceedings of the International Conference on Shape Memory and Superelastic Technologies*; Russel, S.M.; Pelton, A.R.; Eds.; SMST-2000: Fremont, US, 2001; pp 1-9.
[47] Wu, M.H. *Mater. Sci. Forum* 2002, 394-395, 285-292.
[48] Wang, L.M.; Liu, L.H.; Yang, H.; Wang, L.Y.; Xiu, G.Q. *Mater. Sci. Forum* 2002, 394-395, 297-300.
[49] Bondarev, A.B.; Andreyev, V.A. *Mater. Sci. Forum* 2002, 394-395, 301-304.
[50] Khalil-Allafi, J.; Dlouhy, A.; Eggeler, G. *Acta Mater.* 2002, 50,4255-4274.
[51] Frenzel, J.; Neuking, K.; Eggeler, G. *Mat.-wiss. u. Werkstofftech.* 2004, 35, 352-358.
[52] Frenzel, J.; Zhang, Z.H.; Neuking, K.; Eggeler. G. *J. Alloys Comp.* 2004, 385, 214-223.
[53] Zhang, Z.H.; Frenzel, J.; Neuking, K.; Eggeler, G. *Acta Mater.* 2005, 53, 3971-3985.
[54] Ballamy, W.C.; Hucke, E.E. *J. Metals* 1970, 22, 43-50.
[55] Pohl, M.; Heßing, C.; Frenzel, J. *Mater. Sci. Eng. A* 2004, 378, 191.

[56] Massalski, T.B. Ed.; *Binary Alloy Phase Diagrams*; 2nd Ed.; ASM International: Ohio, Materials Park, US, 1990; Vol.1, pp 2874-2875.
[57] Du, Y.; Schuster, J.C. *Z. Metallkd.* 1998, 89, 399-410.
[58] Stover, E.R.; Wulff, J. *Trans. Met. Soc. AIME* 1959, 215, 127.
[59] Olier, P.; Barcelo, F.; Bechade, J.L.; Brachet, J.C.; Lefevre, E.; Guenin, G. *J. Phys. IV France* 1997, 7, C5-143-148.
[60] Kurz, W.; Fisher, D.J. *Fundamentals of Solidification*; 3rd Ed.; Trans Tech Publications Ltd.: Aedermannsdorf, CH, 1989; pp 21-292.
[61] Zener, C. *Trans. AIME,* 1946, 167, 550.
[62] Zhong, W.Z.; Hua, S.K. *Geometry of Crystal Growth*; Science Press: Beijing, CN, 1999; pp 205-208.

In: Trends in Crystal Growth Research
Editor: George V. Karas, pp. 101-132

ISBN 1-59454-541-3


*Chapter 5*

# BRIDGMAN GROWTH OF OXIDE SINGLE CRYSTALS FOR PIEZOELECTRIC APPLICATIONS

***Jiayue Xu and Shiji Fan***
Shanghai Institute of Ceramics, Chinese Academy of Sciences, 1295 Dingxi Road, Shanghai 200050, People's Republic of China

## ABSTRACT

Oxide single crystals are usually grown by Czochralski method. However, the vertical Bridgman method has become an important technique for industrial growth of oxide crystals in recent years. The vertical Bridgman method has several advantages for growing piezoelectric crystals. First, the crystal shape depends on the shape of the crucible; which means high utility of as-grown crystals during processing standard wafers. Second, the multi-crucible Bridgman furnace has been developed for mass-production of piezoelectric oxide crystals with lower cost. Third, a small temperature gradient in the solid-liquid interface and in situ annealing process considerably reduce the thermal stress in as-grown crystals. Lithium tetraborate crystal $Li_2B_4O_7$ (LBO) is a superior substrate material for surface acoustic wave (SAW) and bulk acoustic wave (BAW) devices. The Bridgman growth of 3 and 4 inches LBO crystals is presented in detail and mass production of large size LBO piezocrystals has been realized in our institute. Novel piezoelectric $Sr_3Ga_2Ge_4O_{14}$ (SGG) crystal up to 2 inches in diameter has been grown successfully by the modified vertical Bridgman method for fabrication of SAW devices. Relaxor ferroelectric crystals, such as, $(1\text{-}x)Pb(Mg_{1/3}Nb_{2/3})O_3\text{-}xPbTiO_3$ (PMN-PT) and $(1\text{-}x)Pb(Zn_{1/3}Nb_{2/3})O_3\text{-}xPbTiO_3$ (PZN-PT), possess an extremely large electric-field-induced strain (>1%) and a very high electromechanical coupling coefficient $k_{33}$ (>90%), which make them the next-generation electromechanical transducer materials in advanced applications of actuators, medical ultrasound imaging and underwater communication. The compositional inhomogeneity of crystal wafers and the erosion of high temperature solution on platinum crucible are the main problems in Bridgman growth of these lead-based ferroelectric perovskite crystals. PbO flux has been induced to Bridgman growth of PZN-PT crystals and several auxiliary techniques have been developed to improved crystal quality.

# 1. Introduction

The Bridgman method is generally characterized by the relative translation of the crucible containing the melt to the axial temperature gradient in a vertical furnace. It is very suitable to grow congruently or near congruently melting compounds [1]. Since P.W. Bridgman used a vertically mounted tubular electric furnace to grow some low-melting point metals based on directional solidification of the melts in 1925, the vertical Bridgman method has become an important technique for crystal growth of metals, halides and oxides. A sophisticated modification of the vertical Bridgman furnace was made to grow LiF single crystal by D.C. Stockbarger in 1936 [2]. The modified Bridgman method has been popularly applied to grow II -VI compounds for thermo-electric and semiconductor applications [3]. Due to its simple and effectiveness, the vertical Bridgman method was employed to grow mica crystals in Shanghai Institute of Ceramics, Chinese Academy of Sciences, in the early of 1960's. Many scientists had made their contributions to this work and booklet-like mica crystals up to 110mm × 200mm × 20mm were finally grown in the late 1970's. From then on, more than 30 oxide crystals have been grown successfully by the modified vertical Bridgman method in the institute [4]. Many oxide crystals, such as $Li_2B_4O_7$, $Sr_3Ga_2Ge_4O_{14}$ and $LiNbO_3$, were investigated for piezoelectric applications. There are several advantages for the Bridgman growth of piezoelectric oxide crystals. First, the crystal shape depends on the shape of the crucible; which means high utility of as-grown crystals during processing standard wafers. Second, the multi-crucible Bridgman furnace has been developed for mass-production of piezoelectric oxide crystals. Third, a small temperature gradient in the solid-liquid interface and in situ annealing process considerably reduce the thermal stress in as-grown crystals. Lastly, non-vacuum growth of these oxide crystals saves the cost of the furnace. In this chapter, the growth of oxide crystals are discussed in detailed and the developments of some related techniques are introduced.

# 2. Growth of Large Size $Li_2B_4O_7$ Crystals

## 2.1. Multi-Crucible Growth of $Li_2B_4O_7$ Crystals

Single crystal of lithium tetraborate $Li_2B_4O_7$ (LBO) is a new non-ferroelectric piezoelectric material that has a tetragonal structure with lattice parameters a = b = 9.47 Å and c = 10.26 Å. LBO single crystal is a superior substrate for surface acoustic wave (SAW) and bulk acoustic wave (BAW) devices since it has a specific crystallographic plane (110), which provides a zero temperature coefficient of frequency and a fairly high electromechanical coupling factor ($k^2$ = 1%) [5]. The acoustic velocities of LBO crystals are slightly greater than $LiNbO_3$ and $LiTaO_3$ while the SAW reflectivity per stripe is 5 times greater than $LiNbO_3$, $LiTaO_3$, or quartz, which suggests miniaturization of SAW devices [6]. Other advantages include abundant availability of raw materials and no environmental pollution.

Since Robertson et al reported the Czochralski growth of LBO single crystal in 1982 [7], many efforts had been made to produce large size LBO crystals [8-9]. Macro defects, such as cores, cracks, twins and striations, are often encountered due to the high viscosity of the melt and the coexistence of [$BO_4$] tetrahedral and [$BO_3$] triangles in the framework structures of $Li_2B_4O_7$. These problems seriously hindered industrial application of LBO crystals. Accordingly, the modified vertical Bridgman method was employed to grow large size LBO crystals in 1987 and 3-inch LBO crystals had been grown successfully in 1990 in our laboratory [10].

The Bridgman growth of LBO crystals was carried out in a home-made vertical furnace, as shown schematically in Fig.1. The electric resistance heating system was composed of several $MoSi_2$ heaters. The furnace temperature was monitored by DWT-702 with a proportional integral differential (PID) controller. The vertical temperature distribution in the furnace was generally divided into three zones: the high-temperature zone, the temperature gradient zone and the low-temperature zone, as shown in Fig.2. The high-temperature zone was kept at 40~80°C higher than the melting point of LBO crystal and polycrystalline LBO powder was allowed to have enough time for melting and homogenization in the enclosed crucible. A temperature gradient of 20~40°C/cm was formed in the gradient zone and the top of the seed was melted in this zone for seeding. The length of the low-temperature zone was about 20cm with a temperature range of 700-850°C. This zone acts as an in-situ annealing system after the growth process finished. The thermal stress in as-grown crystals was reduced as little as possible by careful control of the temperature in this zone. Due to high viscosity of the melt and lack of the forced convection in the melt, the lowering rate is very slow (less than 0.3mm/h). In order to increase the yield, a multi-crucible furnace has designed. Generally, 5~8 crucible placements for 3-inch LBO and 3~6 for 4-inch LBO are arranged in the furnace so that several LBO crystals can be grown at the same time.

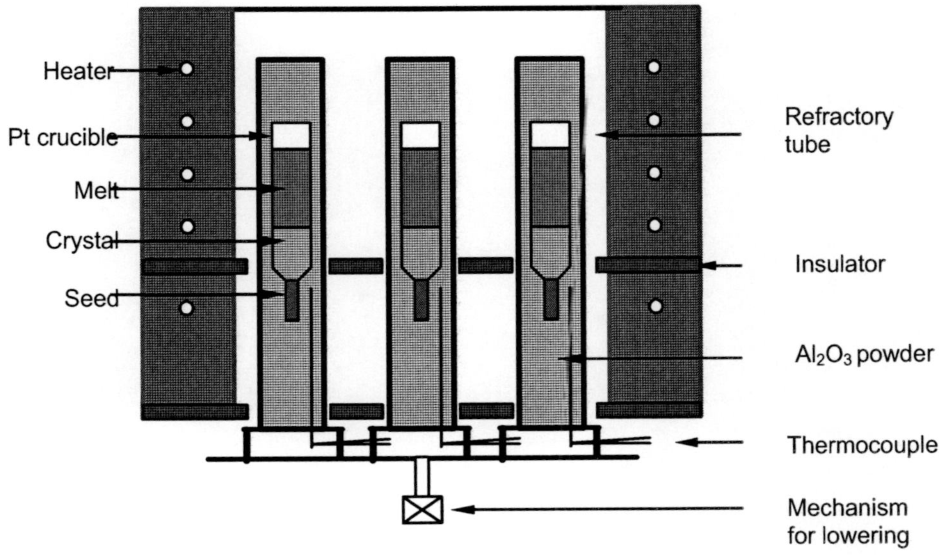


Figure 1. Schematic drawing of the multi-crucible vertical Bridgman furnace

Polycrystalline LBO powder was synthesized by dissolving the mixture of $H_3BO_3$ and LiOH, crystallizing from the solution and dehydrating at high temperature. Slightly $B_2O_3$-rich raw material was necessary for successful growth of high quality LBO crystals. The powder was pressed into round blocks with a density of 1~1.4g/cm$^3$. They were loaded into a platinum crucible and <110>-oriented LBO seed was inserted into the seed well of the crucible in advance. The crucible was moved to a refractory tube, which was filled with $Al_2O_3$ powder to support Pt crucible as well as to isolate it from external temperature fluctuations. A pair of thermocouples was installed in the tube and the temperature at the top of the seed was measured so that the seeding process was achieved according to the value of the thermocouples. In addition to the lowering mechanism, a fine mechanical adjustment system was assembled to each refractory tube so that the height of each crucible in the furnace can be independently manipulated for successful seeding. When the growth process finished, the crucibles were moved slowly to the lower temperature zone and kept more than 10h for annealing, and then cooled to room temperature at a rate of 40°C /h. The growth conditions for multi-crucible growth of 4-inch LBO crystals have been discussed in detail in reference [11].

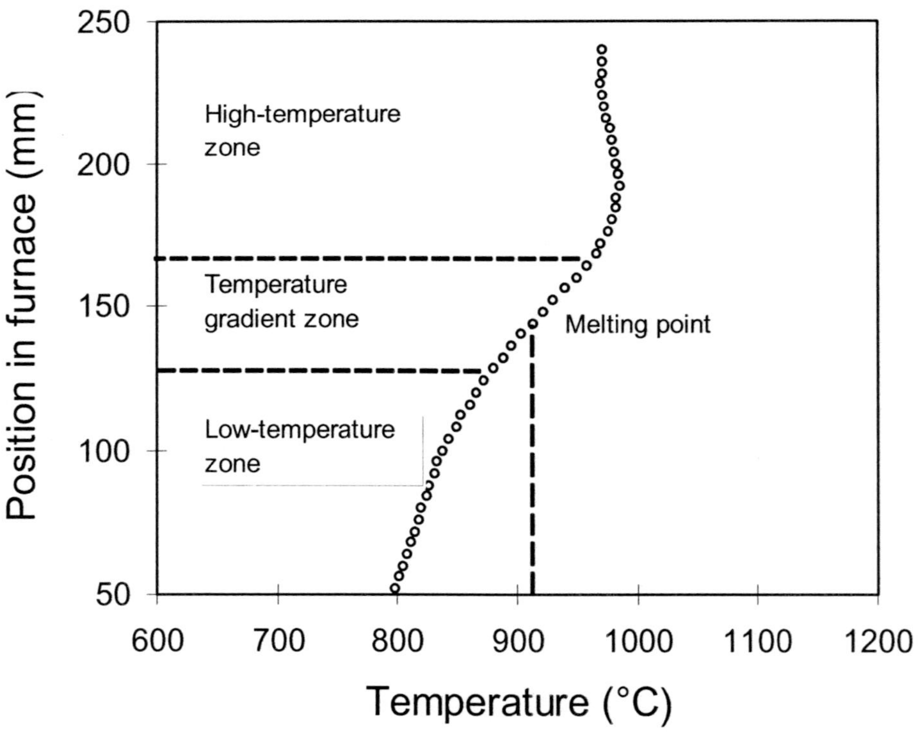


Figure 2. The vertical temperature profile of the Bridgman furnace

Due to the slow growth rate, the growth cycle may continue several months for a long crystal. There is more or less residual melt frozen in the top of the crystal boule because of component deviation. The residual melt is easy to induce LBO crystal cracking. So the tail of as-grown crystal should be cut as soon as possible when as-grown LBO boules are taken from the crucibles. Fig.3 shows a Bridgman-grown LBO crystal boule with a dimension of 82mm diameter and 210mm length. Macro defects, such as cores, striations and cracks, are easily observed by naked eyes due to the good transmittance of as-grown crystals.

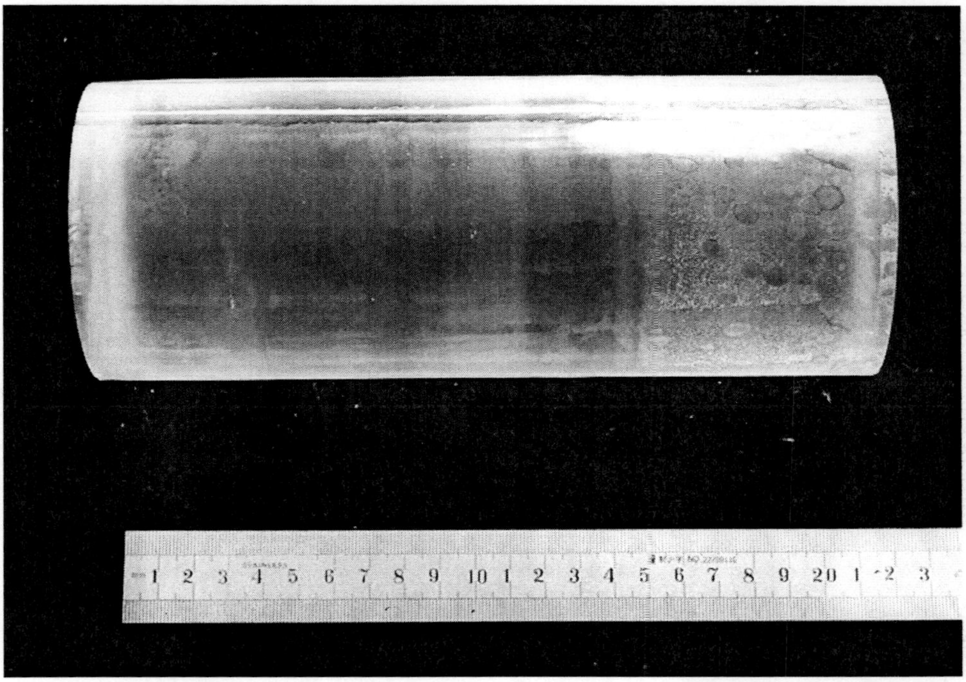

Figure 3. Bridgman-grown LBO crystal boule

## 2.2. Growth Defects and Control of Crystal Quality

Although a very slow lowering rate was adopted, some macro-defects often occurred during the Bridgman growth of LBO crystals. These defects included cores, cloud-like striations, bubbles, twins and cracks. The defects decreased the quality and yield of LBO crystals. Therefore, it is highly significant to investigate growth defects.

Core is a continuous or intermittent opaque region parallel to the growth direction. Generally, the core occurs at the center of the boule, as shown in Fig.4. However, it sometimes appears near the side because of the asymmetrical temperature distribution. In Fig.4, the core diverged the center gradually in the upper part of the boule and disappeared in the fast-cooling region. The mechanism of coring is essentially dendritic growth with growing rifts and constitutional super-cooling inclusions at a concave solid-liquid (S-L) interface. When the shape of S-L interface is not regular symmetrical, the core occurs inclining to the side of LBO boule. Rich Li component and some impurities are found in the core region by means of chemical analysis. A stable temperature profile, a slow growth rate and polycrystalline raw materials with a slight rich $B_2O_3$ are necessary for the Bridgman growth of core-free LBO crystals.

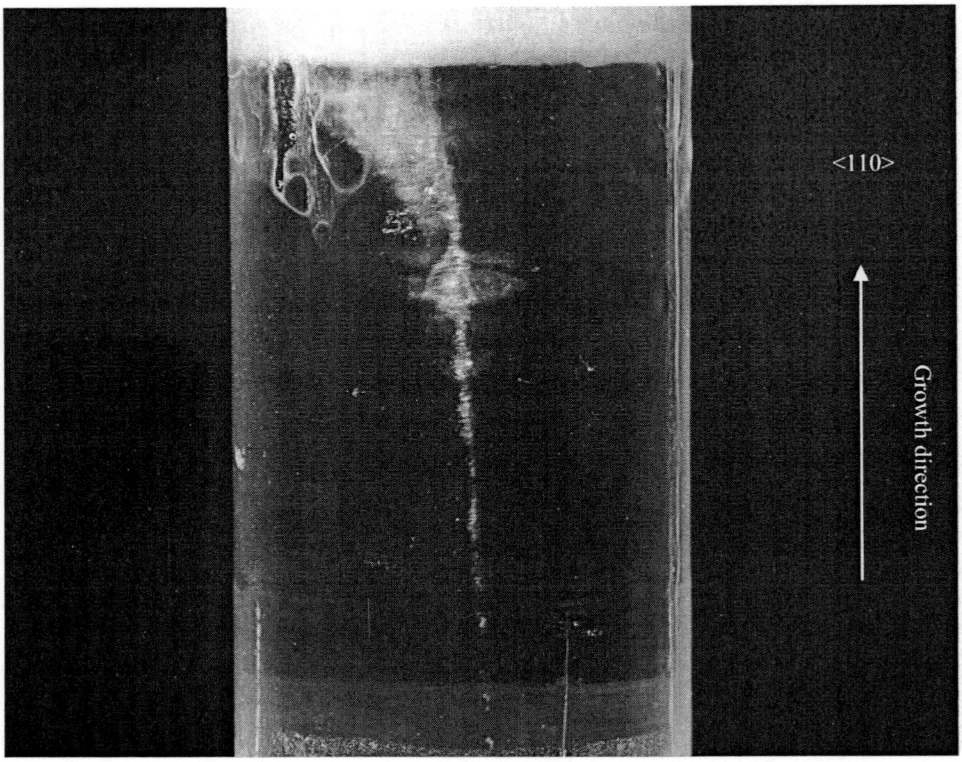


Figure 4. The cores in a Φ3″ LBO boule

The striations are opaque and perpendicular to the growth direction. The temperature fluctuation near S-L interface may result in forming striations. For example, striations are formed in all LBO boules when the environmental temperature or the electric power changes suddenly, as shown in Fig.5. The striations in different boules appeared at the same growth length. Sometimes, the cloud-like striations occur as the symbiosis of the cores. Fig.6 shows the typical striations in a Φ3″ LBO boule. In the middle of the boule, there are cores extended along the growth direction. These cores penetrate the parallel striations. The shape of the striations is mainly concave to the melt. This suggests that the striations and cores are easily formed when the solid-liquid interface is concave. When the furnace temperature fluctuates irregularly, the constitutional supercooling may occur at the bottom of the concave interface, which resulted in the forming of striations and cores [10, 11]. A flat or slight convex interface is helpful for Bridgman growth of high quality LBO crystals. The critical issue is how to control a flat or slight convex interface during Bridgman growth. Due to large heat capacity of the melt compared to that of the seed, the concave interface is easily formed in the initial growth stage. It was found that the interface position shifted upward due to growth acceleration and the shape of S-L interface changed from concave to flat even convex during the growth process [11]. This means that a flat or slight convex interface might be realized by adjusting the furnace temperature and the seeding position. For the growth of 4-inch LBO crystals, the furnace temperature was optimized as 40~80°C higher than the melting point of LBO crystal and the temperature gradient at S-L interface was modified as about 20-40°C/cm [11].

Figure 5 Striations caused by sudden temperature fluctuations

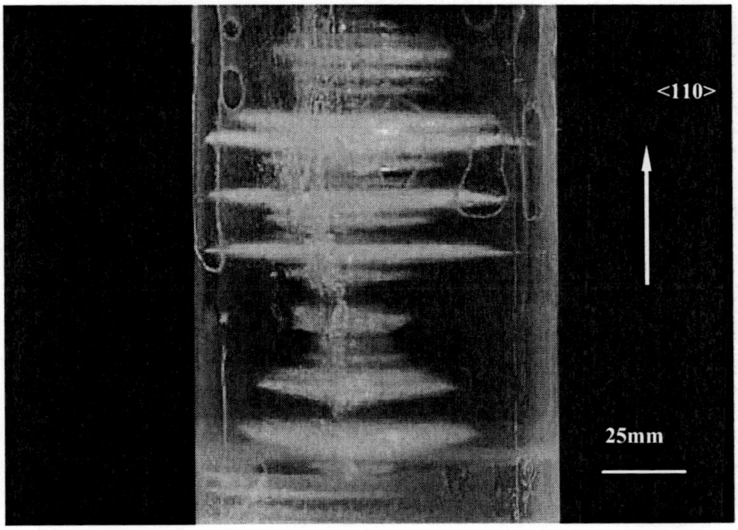


Figure 6. The typical striations in a Φ3″ LBO boule

As described in the above section, the raw materials of LBO polycrystalline powders are melted gradually for seeding when the crucible goes up. A stable thermal equilibrium should be achieved near the top of the seed and crystal is grown along the seed. Due to high viscosity of the melt and absence of forced convection in the melt during Bridgman growth, small amount of air or moisture in the melt moved slowly from the solid-liquid interface to the upper part of the crucible. As a result, some bubbles are frozen in the crystal during the initial stage of the growth. Fig.7 shows two lines of small bubbles. Generally, the bubbles are

0.4~1mm in diameter and each line of the bubbles may continue 5~16mm. As mentioned above, polycrystalline LBO raw material was prepared from the solution. Preparation of anhydrous LBO powder is a key step for LBO crystal growth. Moreover, some measures are taken to eliminate bubbles, for example, increasing the furnace temperature and holding the melt longer time.

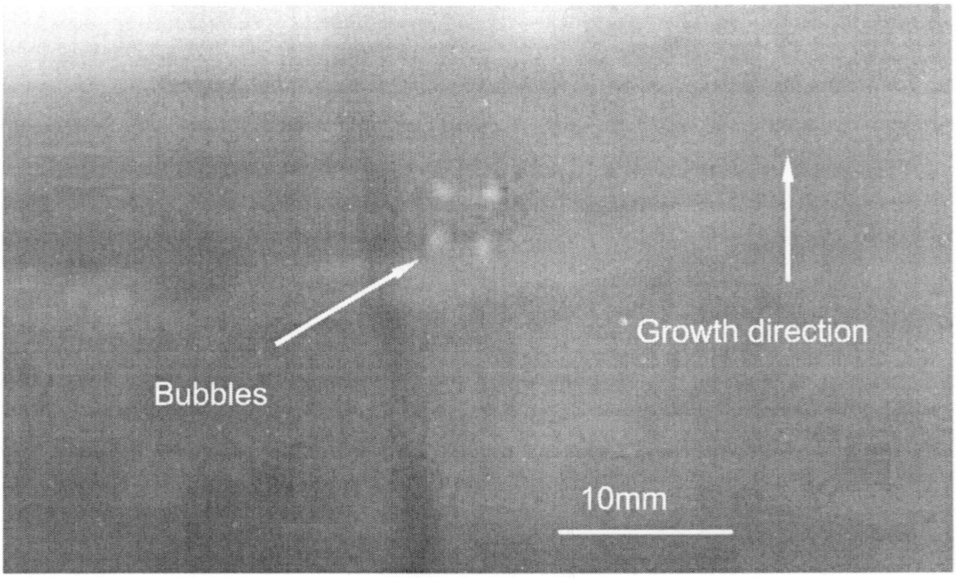


Figure 7. Small bubbles in the Bridgman-grown LBO crystal

The coexistence of $[BO_4]$ tetrahedral and $[BO_3]$ triangles in the framework structures of $Li_2B_4O_7$ makes the crystal cracking easily. So as-grown crystal cannot bear large thermal stress. This is the main reason that brought failure in Czochralski growth of large size LBO crystals. Besides, growth defects can induce cracking also. Fig.8 shows the cracking along twin boundary. During the Bridgman growth, the cracking is improved considerably because of its smaller temperature gradient at the S-L interface (~20-40°C/cm for 4-inch LBO). However, the existence of the residual melt on the top of as-grown boule is very harmful and usually induces cracking due to the strong anisotropy of thermal expansion. The crystals with different growing orientations show different cracking behaviors. For <110>-oriented LBO crystals, cracks often occur along (100) and (010) faces symmetrically. For <001>-oriented crystals, cracking occurs frequently along (110). The cracks induced by the residual usually extend about 30mm, even longer. It is found that the raw materials with a slight excess of $B_2O_3$ are helpful for the growth [10]. The growth experiments demonstrate that it is better to control the weight ratio of $Li_2O$ and $B_2O_3$ in the range of 0.2125~0.2146. The upper limit of the value is the weight ratio of $Li_2O$ and $B_2O_3$ in stoichiometrical $Li_2B_4O_7$. Furthermore, an in-situ annealing process has been employed to reduce the thermal stress in as-grown crystal. The process is to move the crucible to the lower-temperature zone and held it at 600~800 °C for 24h, and then cool down slowly at a small rate of 20-30 °C/h. Through this process, thermal stress is released from as-grown crystal, especially from the upper part of the boule, and the occurrence of cracking is reduced considerably.

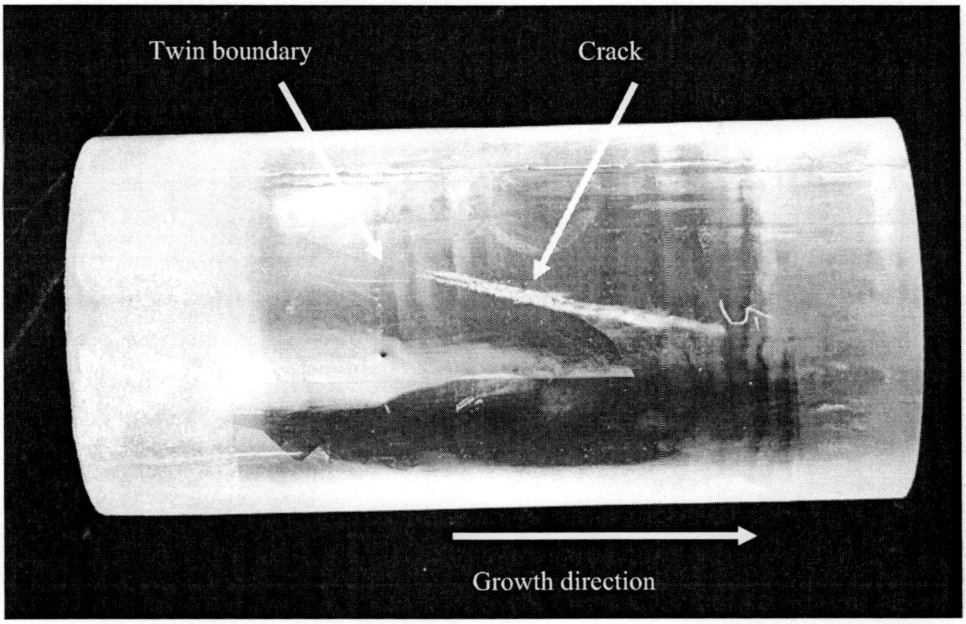


Figure 8. The crack occurring along the twin boundary

Twins in LBO crystals are demonstrated by the etching method with 25% acetic acid for one hour at room temperature. Fig.9 shows a twin near the side of LBO boule. Twins appear in the side may be induced by the powder particles between seed and crucible wall or the protrusion on the inner wall of Pt crucible. It is difficult to make a completely smooth inner wall of the crucible, especially in the shouldering place. So twins are easy to originate from the protrusion on the inner wall of Pt crucible. Twins are also observed propagated from the seed defects, such as striation, core, crack or small original twins. The twins on (110) plane of LBO boules display obvious crystallographic habits although the shapes of the twins on (110) plane may be different. These twins usually appear as polygons with a same inner angle. For example, the inner angle of the boundary A and B in Fig.9 is about 76°. This result suggests that these twin boundaries are related to some crystallographic planes. Some twins disappear gradually along the growth direction, as shown in Fig.8, while others may become more and more large. Sometimes, several twins originate at different points of the side in one boule. They may grow larger and combine together at last.

An unusual twinning phenomenon was found in LBO crystals. Several twins appeared parallel to the growth direction, as shown in Fig.10. These laminar twins often occurred in LBO boules grown along [001] or [011] direction. The twinning takes place according to Dauphine rule with twinning axes [100] and [010] in LBO crystals. The laminar twins have been also observed in Czochralski-grown LBO crystals [12]. Considering the polarity of the c-axis and the different etching rates in the twinning and the normal regions, these laminar twins are regarded as electrical twinning and have no effect on optical properties. However, the existence of twins has an effect on piezoelectric and acoustic properties. Twin-free LBO crystals of 3 or 4 inches in diameter have been grown successfully in our laboratory for surface acoustic wave (SAW) applications.

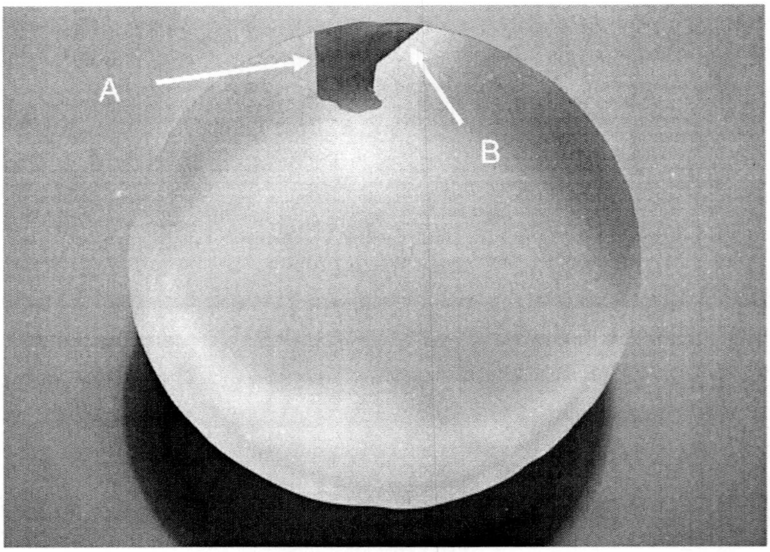


Figure 9. The twin in (110) plane

Figure 10. The laminar twins observed in (011) plane

In order to control crystal quality and achieve industrial growth, the optimizing growth process for Bridgman growth of LBO crystals can be summarized as following: (1) Considering the high viscosity of the melt, the growth rate is very slow, usually less than 3mm/day, and a flat or slightly convex S-L interface should be achieved; (2) The temperature gradient at S-L interface is about 20~40°C/cm and an in-situ annealing process is adopted to reduce the cracking; (3) Polycrystalline LBO powder is synthesized and the weight ratio of $Li_2O$ and $B_2O_3$ is controlled in the range of 0.2125~0.2146; (4) The slow growing inevitably results in low yield of LBO crystal, so multi-crucible Bridgman furnace is developed. By optimizing the growth parameters, striation-free bubble-free, twin-free and crack-free LBO crystals up to 105mm in diameter have been grown reproducibly. Fig.11 shows some boules of 3-inch and 4-inch LBO crystals.

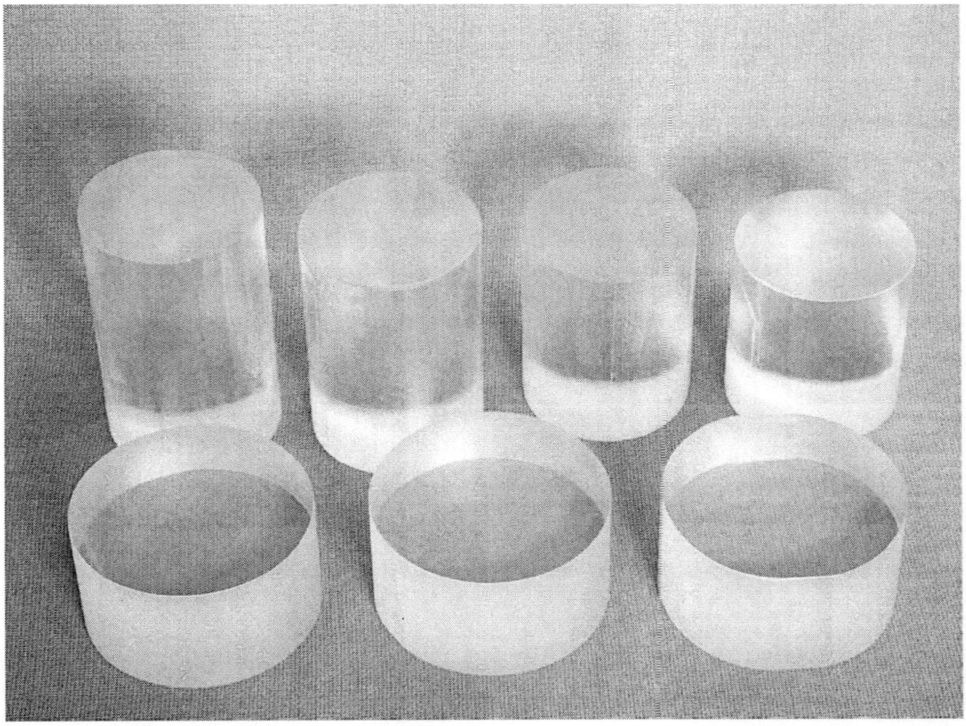

Figure 11. Some boules of 3-inch and 4-inch LBO crystals

## 2.3. Piezoelectric Properties and SAW Applications

LBO is a piezoelectric crystal with the 4mm point group, which has six elastic, three piezoelectric and two dielectric independent components. These piezoelectric constants have been measured at room temperature. Other properties, such as SAW wave velocity, electromechanical coupling factor, thermal expansion coefficients and mechanical strengths, are very important for SAW application and are measured also. Table 1 gives some physical properties of Bridgman-grown LBO crystals. LBO is a multi-functional crystal, so its thermoelectric, electro-optic and acousto-optic coefficients are also listed in Table 1.

As a superior piezoelectric substrate, LBO is generally used to fabricate SAW resonator and resonator filter, especially suitable for SAW resonator filters which have the relative bandwidth of 0.3% to 0.6% considering coupling coefficient and temperature coefficient of delay [17]. Also, the chip size of SAW resonator filters can be reduced because of the large reflection coefficient for aluminum strips on LBO substrates. These SAW resonator filters show the advantages of low insertion loss, high rejection level and small size, and are extensively used in GSM, cordless telephone receiver and other applications [17, 18]. Φ3″ LBO crystals have been mainly supplied to some Japanese companies for fabrication of SAW devices. Recently, some Chinese companies show great interests in LBO substrates due to fast development of mobile communications.

**Table 1. Physical properties of LBO crystals**

| | | | | References |
|---|---|---|---|---|
| Piezoelectric strain const. ($10^{-12}$ C·N$^{-1}$) | $d_{13}$ =8.19 | $d_{31}$ =-2.63 | $d_{33}$ =19.5 | [13] |
| Piezoelectric stress const. (C·m$^{-2}$) | $e_{13}$ =0.46 | $e_{31}$ =0.23 | $e_{33}$ =0.91 | [13] |
| Elastic stiffness const. ($10^{10}$ N·m$^{-2}$) | $C^E_{11}$ =12.84 | $C^E_{12}$ =0.46 | $C^E_{13}$ =2.97 | [13] |
| | $C^E_{33}$ =5.42 | $C^E_{44}$ =5.78 | $C^E_{66}$ =4.75 | [13] |
| Elastic compliance const. ($10^{-12}$ m$^2$·N$^{-1}$) | $S^E_{11}$ =8.82 | $S^E_{12}$ =1.0 | $S^E_{13}$ =-5.5 | [13] |
| | $S^E_{33}$ =24.3 | $S^E_{44}$ =17.3 | $S^E_{66}$ =21.1 | [13] |
| Dielectric const. | $\varepsilon^T_{11}/\varepsilon_0$ =9.33 | $\varepsilon^T_{33}/\varepsilon_0$ =10.1 | | [13] |
| | $\varepsilon^S_{11}/\varepsilon_0$ =8.91 | $\varepsilon^S_{33}/\varepsilon_0$ =8.07 | | [13] |
| Electromechanical coupling factor | $k^2$ = 1% (45° X-cut and Z propagation) | | | [13] |
| SAW wave velocity (m/s) | 3873±21 (Z-cut, X-propagation) | 3510±6 (X-cut, Z-propagation) | 3401 (45°X-cut, Z-propagation) | [14] |
| Thermoelectric const. (μC·m$^{-2}$·K$^{-1}$) | $p_T$ = − 30 (25℃) | | $p_T$ = − 120 (-150℃) | [13] |
| Electro-optic coefficient ($10^{-12}$ m·V$^{-1}$) | $\gamma_{13}$ =2.64 | $\gamma_{33}$ =2.55 | $\gamma_{51}$ =−1.24 | [15] |
| Acousto-optic coefficient ($10^{-18}$ S$^3$·g$^{-1}$) | $M_{13}$ =0.23 | $M_{33}$ =0.28 | | [16] |
| Compressive strength (Kg·cm$^{-2}$) | 2878 (‖<100>) | 5750 (‖<001>) | 4496 (‖<110>) | [13] |
| Bending strength (Kg·cm$^{-2}$) | 1946 (⊥<100>) | 950 (⊥<001>) | 1886 (⊥<110>) | [13] |
| Thermal expansion coefficient ($10^{-6}$/°C) | $\alpha_{11}$=11 (20-35°C) | | $\alpha_{33}$=3.7 (20-35°C) | [13], [16] |

# 3. Growth of $Sr_3Ga_2Ge_4O_{14}$, $LiNbO_3$, $Bi_{12}Si(Ge)O_{20}$ and $TeO_2$ Crystals

## 3.1. Growth of $Sr_3Ga_2Ge_4O_{14}$ Crystals

Langasite ($La_3Ga_5SiO_{14}$: LGS) single crystal, with a zero temperature coefficient for the resonant frequency at room temperature and an electromechanical coupling coefficient which is three times larger than that of quartz, has been considered to be a promising piezoelectric material for bulk acoustic wave (BAW) and surface acoustic wave (SAW) devices [19]. Its property of slow acoustic velocity makes it possible to fabricate small size intermediate frequency SAW filter in mobile communication system, wideband-code division multiple access (WCDMA) [20, 21]. However, large size LGS wafers are very expensive due to growth difficulties and high price of the raw materials. Langasite has a trigonal structure, which belongs to point group 32 and space group P321. By aliovalent substitutions, a large number of isostructural compounds have been prepared. Among them, $Sr_3Ga_2Ge_4O_{14}$ (SGG) crystal attracted our attention due to its higher piezoelectric constants and lower melting temperature compared with LGS crystal [21].

SGG crystal has been grown by Czochralski method for piezoelectric applications [22]. The crystal was small and the growth of high quality SGG crystal by the Czochralski method remains technically difficult due to the strong evaporation of $GeO_2$ from gallogermanate melts. In Bridgman growth of SGG crystals, SGG polycrystalline powder is pre-synthesized from high purity $SrCO_3$, $Ga_2O_3$ and $GeO_2$ by sintering the mixture at 1200°C for 16h. Due to lack of SGG seeds, a LGS crystal of Φ10mm × 40mm was used as seed. As a hetero crystal seed, LGS has proved to be a good choice for Bridgman growth of SGG crystals because the higher melting temperature makes the seeding easy and the same structure makes the lattice mismatch negligible. Fig.12 shows as-grown SGG crystal grown by Bridgman method using a LGS seed. The length of the LGS seed remained almost unchanged while the color weakened compared with the original LGS. The change in LGS color may be attributed to oxygen deficiency caused by the sealed platinum crucible during growth. To make sure whether LGS seed pollutes as-grown crystal, the chemical composition of the SGG crystal has been analyzed by X-ray fluorescence spectra. Fig.13 shows the concentrations of $La^{3+}$ and $Si^{4+}$ ions varied with the growth length. The concentrations of $La_2O_3$ and $SiO_2$ decrease quickly and less than 0.4wt% $La_2O_3$ and $SiO_2$ are measured. Considering the shouldering length is about 30mm, the influence of hetero-composition LGS seed on crystal quality can be negligible for piezoelectric applications.

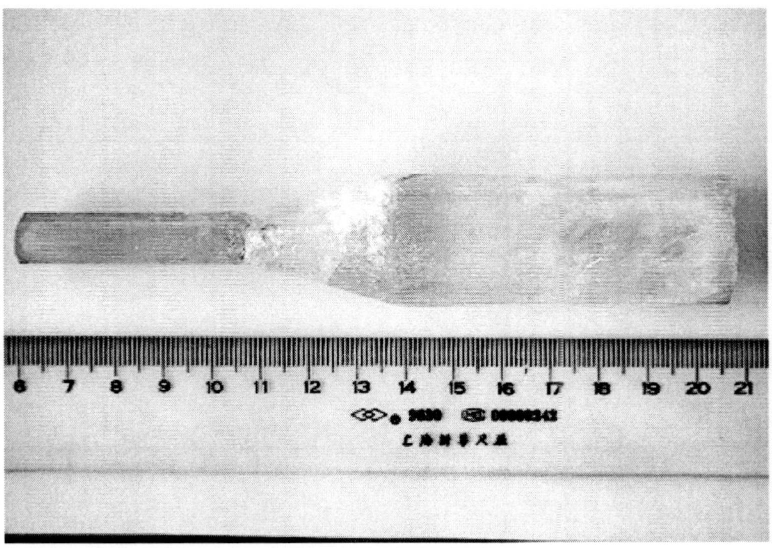


Figure 12. As-grown SGG crystal with LGS seed

Φ2″ SGG crystal has been grown using high-quality and <001>-oriented SGG seed. Due to lack of forced convection in the melt during the Bridgman growth, a suitable temperature gradient and a smaller growth rate are necessary for the successful growth. A large temperature gradient in S-L interface may result in cracking, especially at the shoulder part and the top of as-grown SGG boule. Sometimes, scattering particles are also found in these regions by laser beam. These scattering particles are small inclusions with different morphology. Fig.14 shows some typical inclusions in SGG crystals. The growth results show that the raw materials with 1% and 2% excess of $Ga_2O_3$ and $GeO_2$ are necessary for successful growth of 2-inch SGG [23]. The excessive $GeO_2$ and $Ga_2O_3$ can compensate the composition loss during the sintering and growth process, thus be helpful for the growth of SGG crystals.

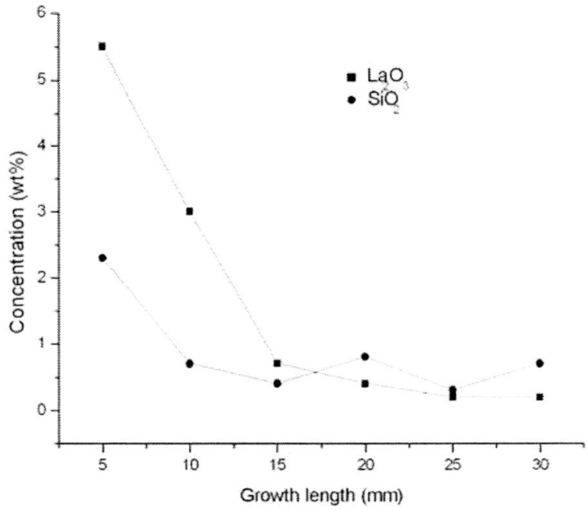


Figure 13. The concentrations of La2O3 and SiO2 vs the growth length

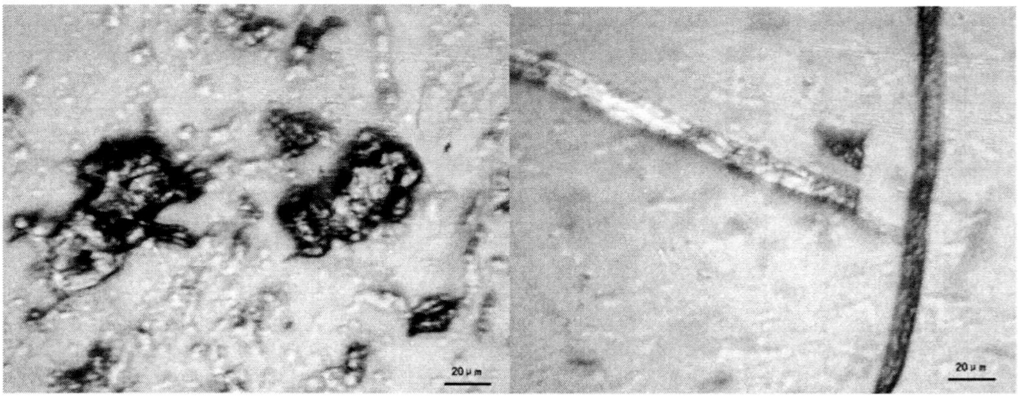


Figure 14. Some typical inclusions in SGG crystals

Crack-free and inclusion-free transparent SGG crystal of 2 inches in diameter has been grown by the modified Bridgman method, as shown in Fig.15. Its piezoelectric constants have been obtained by measuring resonance-antiresonance frequencies using the impedance analyzer. Two independent components $d_{11}$ and $d_{14}$ are $-7.41\times10^{-12}$C/N and $7.05\times10^{-12}$C/N, respectively. The results are slightly larger than that of LGS.

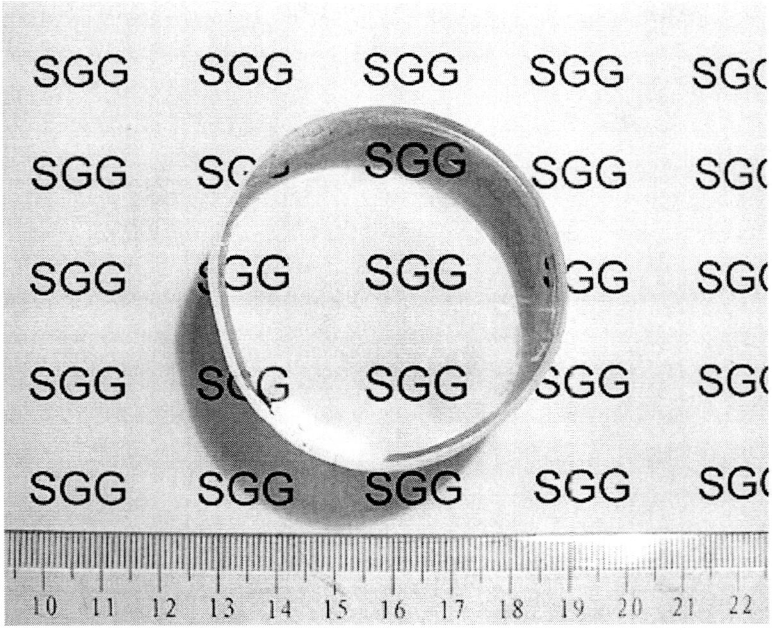


Figure 15. Φ2″ SGG crystal grown by the modified Bridgman method

## 3.2. Growth of Linbo$_3$ Crystals

Lithium niobate ($LiNbO_3$: LN) single crystal has attracted much attention due to its outstanding non-linear, electro-optical, acousto-optical and piezoelectric properties. Pure and doped LN crystals have been extensively used in optical switches, optical modulators and frequency conversion, especially in surface acoustic wave (SAW) and bulk acoustic wave (BAW) devices as an excellent substrate material. Since the first successful growth of the $LiNbO_3$ crystals in 1965 by Czochralski (CZ) method [24], almost all commercially available $LiNbO_3$ crystals have been grown from the melt of congruent composition by this process. Taking account of the congruent characteristics and a moderate melting temperature, we consider that the vertical Bridgman method is another favorable technique for the growth of $LiNbO_3$ crystals.

Using congruent LN seed, LN crystal of 25mm in diameter is easily grown [25]. However, many unexpected problems have been encountered during the Bridgman growth of large size LN crystals. First, LN seed cracks seriously each time due to the existence of ferroelectric domains, which makes it impossible to reuse the seed. Secondly, sparks are often observed when the crucible is torn off. The sparks are dangerous and may induce cracks in as-grown crystal. The sparks may be attributed to the accumulation of pyroelectric charges when the crucible is cooling down. Thirdly, as-grown crystal shows a pale brown color due to lack of oxygen. This may be harmful to its optical application, but it is not so bad for SAW application. It is well known that large amount of surface charges in LN wafers may cause processing or operational failures. The existence of oxygen vacants can improve its electric conductivity. In doing photolithography on LN wafers, its high transparency introduces the problem of diffuse reflections off the wafer's back surface, which decreases the contrast of the exposed resist and causes inconsistent linewidths [26]. Therefore, many efforts have been made to improve its electric conductivity and decrease its optical absorption, such as chemical reduction processing. Furthermore, Bridgman-grown LN crystals are nearly single domain, which means that as-grown LN crystals can be processed into wafers without poling. These growth results demonstrate that the vertical Bridgman method is a suitable technique to grow LN crystals for piezoelectric applications. Fig.16 shows a 2-inch LN crystal grown by Bridgman method. The further work is to develop multi-crucible Bridgman process to grow 3-inch LN crystals for industrial applications.

Figure 16. $LiNbO_3$ crystal grown by the vertical Bridgman method

## 3.3. Growth of $Bi_{12}Si(Ge)O_{20}$ Crystals

Bismuth silicon oxide ($Bi_{12}SiO_{20}$ or BSO) and bismuth germanium oxide ($Bi_{12}GeO_{20}$ or BGO) crystals belong to the rare space group 123 and have a combination of unusual piezoelectric, acousto-optical, photoconductive, photorefractive, and electro-optical properties, which have led to their widespread uses in real-time holography, optical data storage, electro-optic and surface acoustic wave devices [27]. BSO and BGO crystals are congruent compounds and can be grown by Czochralski and Bridgman method [28]. The defects, such as growth striations, darkened cores and strain-induced birefringence are often encountered in Cz-grown crystals. These defects, especially the cores, reduce the quality and availability of as-grown BSO and BGO crystals, and thereby hinder the practical utilization of these crystals [29]. Xu Xuewu et al has investigated the Bridgman growth of BSO crystals for the first time. The darkened cores are observed at the central or off-central part of BSO crystal grown along <001> and <110>, but no darkened cores in <111> and <112>-grown BSO crystals. The similar results have also been found in Cz-grown BSO crystals [30]. It has been suggested that the cores are related to the growth facets [31]. The origin of the cores is ascribed to the preferential substitution of $Bi^{3+}$ ion for $Si^{4+}$ ion in the facets due to the difference in the growth mechanisms on the facets and over the rest of the interface. The pronounced anisotropic growth results in the appearance of (110) and (001) facets on the high convex S-L interface, which shows quite different morphologies for the growth directions <001>, <110>, <111> and <112> [29]. Because the formation of the darkened cores is dependent on the growth direction and interface shapes, the interface convex to the melt can be accomplished and the darkened core can be eliminated and avoided by adjusting the structure of the furnace and the seeding position.

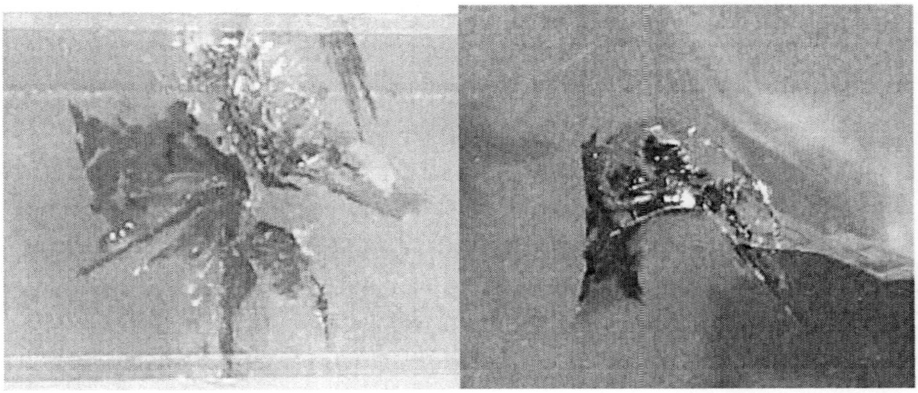

Figure 17. Patterns of Pt-related inclusions

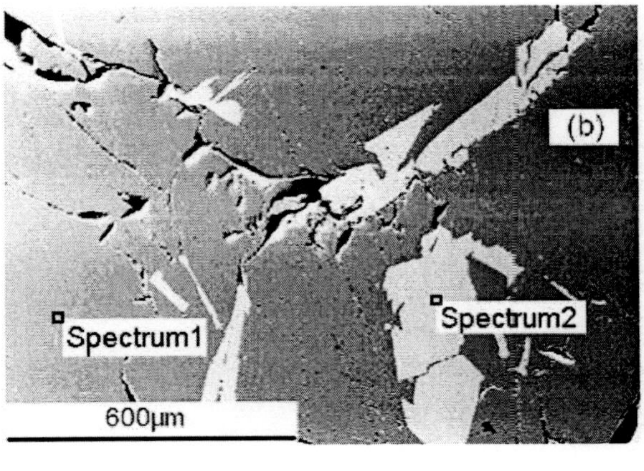


Figure 18. EPMA photos of Pt-related inclusions

Besides, the erosion of the melt on Pt crucible is another troublesome problem in the growth of BSO crystal. The erosion may result in the formation of some Pt-related defects, such as Pt inclusions, particles and cracks, as shown in Fig.17. The defect region is investigated by means of Electron Probe X-ray Microanalysis (EPMA). EPMA results shows that the inclusions appears with irregular shapes, in the form of triangle flakes, strips or saw-tooth pieces, sometimes aggregating into large inclusions with the size of 30μm to 5mm, as shown in Fig.18. In fact, the crucible erosion was a common phenomenon during the growth of BSO and BGO crystals due to the high content of $Bi_2O_3$ in the melts. The contamination of $Bi_2O_3$ impurity makes the reaction of the melt with the Pt crucible more and more serious and results in the inclusions of some separated metallic pieces in as-grown BSO crystals [32]. Lower furnace temperature and purified crucible are necessary to reduce the Pt-related defects. By optimizing growth conditions, core-free and striation-free BSO crystals without cracks and scattering particles have been grown by the modified Bridgman method along

<111> and <112> directions at a growth rate of less than 1mm/h, which provides a possibility of mass production of high quality BSO crystals. Similar problems have been encountered also in the Bridgman growth of BGO crystals. Fig.19 shows large size BSO and BGO crystals grown by the vertical Bridgman method.

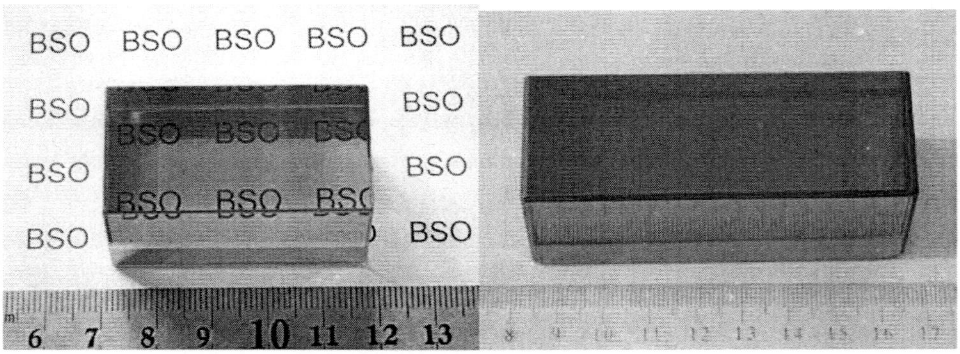


Figure 19. Polished BSO and BGO crystals by the vertical Bridgman method

## 3.4. Growth of $TeO_2$ Crystals

Tetragonal paratellurite ($\alpha$-$TeO_2$) crystal is a compound well known for its attractive acousto-optic properties and is commonly used in various applications such as light deflection/modulation, tunable frequency filtering or Q-switching of laser beams [33]. Besides, $\alpha$-$TeO_2$ is an important piezoelectric material. The piezoelectric tensor consists of only two non-vanishing elements, $e_{14}$ and $e_{25}$, where $e_{14} = -e_{25}$ [34]. The crystal permits unusual piezoelectric applications because its only piezoelectric coefficient is a shear coefficient $e_{14}$ and it generates and detects only shear waves [35].

Single crystal $TeO_2$ was initially grown by Czochralski method and macro-defects, such as bubbles and inclusions, were often observed in grown crystals [36]. The formation of these defects were attributed to the platinum dissolution by the melt at the crucible wall and the impurity of the the starting material [36]. In order to meet the demand of large size and high quality crystals, Bridgman growth of $TeO_2$ crystals has been investigated in our institute since 1980s. Commercial $TeO_2$ raw material is usually synthesized by the reaction of Te and $HNO_3$. Te is easily wrapped in $TeO_2$ raw material and results in the formation of inclusions during the growth. In our experimental, commercial $TeO_2$ is treated with HCl and $NH_3{\cdot}H_2O$ sequentially, and then heated in a muffle to a certain temperature for 12 hours. Raw material of high purity $TeO_2$ is obtained. The Bridgman growth of $TeO_2$ crystals has been carried out in a multi-crucible furnace so that the yield of the crystal is improved considerably. The furnace temperature is set to 50-80°C higher than that of melting point (733°C). To grow high quality $TeO_2$ without macroscopic defects, a flat or slightly convex interface is necessary, which is achieved by adjusting the position of crucible in the furnace. The temperature gradient at solid-liquid interface is optimized to be about 45°C/cm. Single crystals $TeO_2$ up to 58mm ×58mm ×180mm have been grown successfully and industrial growth of $TeO_2$ crystals has realized in our institute. Fig.20 shows a Bridgman-grown $TeO_2$ crystal. Tetragonal $TeO_2$ crystal belongs to the crystal class $D_4$ and (100) and (010) are its cleavage planes. Cracks often occur in the crystals grown along <100> and <010> directions. In order to avoid grown crystal cracking, $TeO_2$ crystals are usually grown along <110>, <001> and other directions. Even though <110> is chosen as a preferable growth direction for industrial growth of $TeO_2$ crystals, cracking along cleavage plane is sometimes observed, as shown in Fig. 21. Fortunately, cracks are controlled effectively when the crystal is carefully treated in the furnace and cooling down at a slow rate [37].

Figure 20. Bridgman grown $TeO_2$ crystal with four polished plane

Figure 21. Crack along the cleavage plane in $TeO_2$ crystal

# 4. GROWTH OF LEAD-BASED FERROELECTRIC CRYSTALS

## 4.1. Growth of $Pb_5Ge_3O_{11}$ Crystals

Single crystal of lead germanate ($Pb_5Ge_3O_{11}$: PGO) has some applications in pyroelectric detectors, information storage and light signal processing, acousto-electronic devices, opto-electronic devices due to its sensitive optical activity, small dielectric constant and a relatively high pyroelectric coefficient in a wide range of temperature [38]. PGO single crystal was previously grown by the Czochralski method [39] and some growth problems, including inclusions, metallic particles and quality deterioration in air, are still unsolved. To our knowledge, Bridgman method is applied to grow PGO crystal for the first time.

Due to lack of seed crystal, PGO crystal is initially grown by spontaneous nucleation in the Bridgman furnace. After a seed crystal up to Φ10mm × 40mm has been obtained, seeded growth of PGO crystal is carried out with <100>-oriented seed. Small quantities of black slag are observed on a certain side surface of as-grown boule, as shown in Fig.22. These parts are taken out and ground into powders. XRD analysis shows that two phases, $Pb_5Ge_3O_{11}$ and $Pb_3GeO_5$, are co-existed in the black slag. Chemical compositions of the powders are also analyzed by means of XRF method, which demonstrates that the compositions of black slag deviate seriously from the stoichiometric compositions of $Pb_5Ge_3O_{11}$ crystal and the $GeO_2$ content was less than that of the nominal composition [38]. Table 2 gives the compositions of the black slag region and the crystal. The formation of $Pb_3GeO_5$ phase is attributed to the incomplete solid-state reaction of $GeO_2$ and PbO during soaking the raw materials. Higher furnace temperature and longer soaking time can eliminate the black slag.

Figure 22. The black slag region on Bridgman-grown PGO boule

The growth defects, such as inclusion, negative crystal and bubble in Bridgman grown PGO crystals have been investigated. Some insoluble particles have been observed, as shown in Fig.23. The inclusions demonstrate cone-like, pillar-like or snake-like morphologies. EPMA results show that the main composition of the particles is PbO. It is known that the melting point of PbO (886°C) and $GeO_2$ (1115°C) are higher than that of PGO (738°C). To eliminate these particles, the mixture of $GeO_2$ and PbO raw materials should react completely at a higher temperature and polycrystalline PGO should be used to grow PGO crystals. Bubbles are also observed in BR-grown PGO crystals. The formation of gas bubbles is attributed to low growth temperature and fast growth rate. By lengthening the soaking time and slowing down the growth rate, the bubbles have enough time to escape from the growth interface and are rarely occurring in the crystals.

**Table 2. Chemical compositions of PGO crystal, black slag and end part of the boule**

| | Theoretical value (wt%) | As-grown crystal (wt%) | Black slag region (wt%) | End part of boule (wt%) |
|---|---|---|---|---|
| PbO | 78.05 | 78.17 | 81.68 | 76.31 |
| $GeO_2$ | 21.95 | 21.83 | 18.32 | 23.69 |

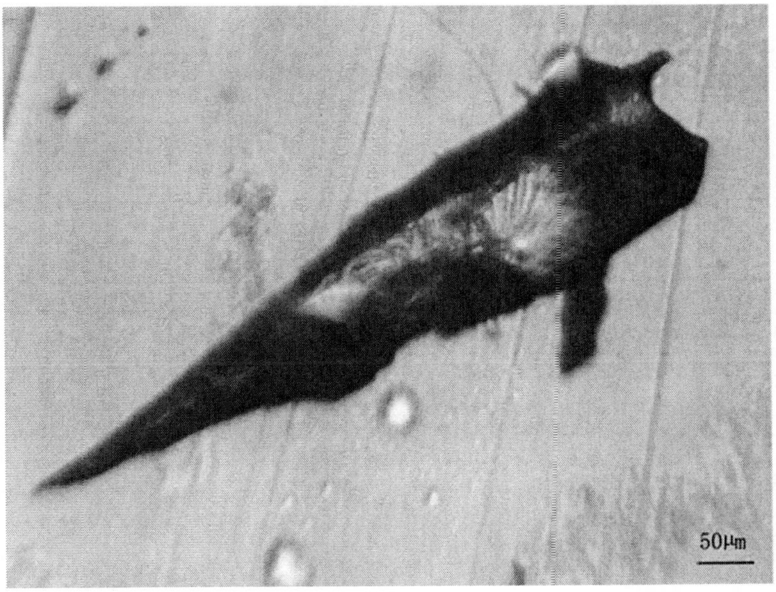


Figure 23. A cone-like inclusion in PGO crystal

The inclusions may induce negative growth in the BR-grown PGO crystals and result in "negative crystal". Fig.24 shows a typical morphology of the hexagonal negative crystal. The sudden temperature fluctuation sometimes occurs because of the instability of heating elements or ambient temperature. In this case, rapid constitutional super-cooling may take place and some melts are wrapped in the crystal. The droplet of the melt then is frozen gradually and forms the negative crystal in the subsequent cooling stage.

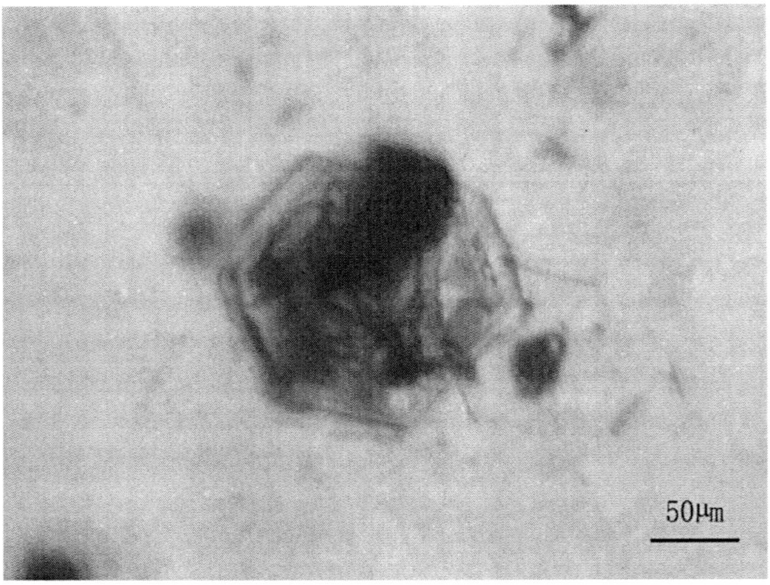


Figure 24. Negative crystal in BR-grown PGO crystal

The temperature dependence of dielectric constants has been measured at the frequencies of 1k, 10k, 100k and 1000kHz using a HP4192A impedance analyzer and the Curie temperature of PGO crystal is about 174°C. Thermal properties have been investigated. The specific heat of the PGO crystals has been measured to be 0.327 J/ (g·K) at 298K. The thermal diffusivity and the thermal conductivity for PGO crystals are about 0.323 $mm^2/s$ and 0.763 W/ (m·K) at 298K, respectively. The average thermal expansion coefficients along a- and c-direction from 323K to 447K are $8.68 \times 10^{-6}$ /K and $6.60 \times 10^{-6}$ /K, respectively.

## 4.2. Growth of Relaxor Ferroelectric PMN-PT Crystals

It is well known that some complex perovskites, $Pb(B'B'')O_3$ (B′=Mg, Zn, Ni, Fe, Sc, In and B″=Nb, Ta and W) so called relaxors which have rhombohedral or monoclinic structures and a tetragonal structure with lead titanate $PbTiO_3$ (PT), exhibit a morphotropic phase boundary (MPB) and show considerably large electromechanical coupling factors and dielectric constants [40]. Among them, much attention has been paid to (1-x)$Pb(Mg_{1/3}Nb_{2/3})O_3$- x$PbTiO_3$ (PMN-PT) and (1-x)$Pb(Zn_{1/3}Nb_{2/3})O_3$ - x$PbTiO_3$ (PZN-PT) crystals.

Solid solution of the (1-x)PMN-xPT system is expected to synergistically combine the functional properties of the relaxor ferroelectrics PMN and the ferroelectric PT. PMN-PT crystal with MPB compositions (x = 0.33-0.35) exhibits very high piezoelectric coefficients ($d_{33}$>2000 pC/N), extremely large piezoelectric strain (1.7%), and very high electromechanical coupling factors ($k_{33}$>90%) [41-42]. Such excellent properties point to a potential revolution in electromechanical transduction for a broad range of applications, such as medical ultrasound imaging and diagnosing, undersea sonar and wireless telecommunication, making the relaxor-based piezocrystals the next generation of transducer materials [43]. PMN-PT crystals are usually grown by the flux method [41]. However, it is difficult to obtain large size and high quality PMN-PT crystals due to the complex compositions and strong evaporation of PbO at high temperatures. Many efforts have been made to grow PMN-PT crystals and the vertical Bridgman technique is regarded as a potential approach to grow large size PMN-PT crystals for industrial applications [44]. So far, PMN-PT boules up to 50mm in diameter and 80mm in length have been grown directly from PMN-PT melt in our institute, as shown in Fig.25.

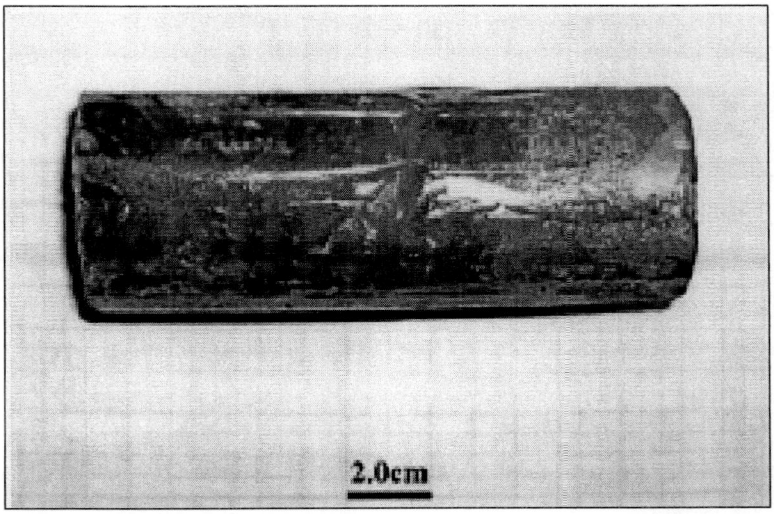


Figure 25. PMNT crystal boule grown by the Bridgman method

The phase diagram of the (1-x)PMN-xPT binary systems has been investigated and the segregation coefficient $k=C_S/C_L$ is less than 1 [45]. Thus, the $PbTiO_3$ content at the initial part of the boule will be less than that in the melt when grown directly from melt, which means obvious chemical and physical inhomogeneity in as-grown crystal. Worse, it is difficult to grow PMN-PT crystal along <001> direction and <001>-oriented wafers are cut sloping to growth direction. This means that even single wafer is inhomogeneous. For example, the Curie temperature $T_C$ varies from 144.4°C at the initial part of the boule to 172.8°C at the ending part and the rhombohedral/tetragonal phase transition temperature changes from from 46.2°C to 88.6°C [45]. Zawilski et al has investigated the Bridgman growth of PMN-PT crystals from initial melts with compositions of 10% PT, 32% PT, and 35% PT. It is found that axial composition is uniform in boules grown from 10% PT melts, but varies from 30–72% PT in a boule grown from a 35% PT. The effective segregation coefficient of $k_{Ti}$ = 0.84±0.04 has been calculated with melts of 32% PT and 35% PT compositions [46]. The results show that segregation is an important factor in the composition region near the morphotropic phase boundary (x = 0.33-0.35). In fact, the piezoelectric properties change dramatically with composition in this region [46].

Owing to the composition complexity, segregation and volatilization of PbO component, a lot of growth defects have been observed in PMN-PT crystals grown by the Bridgman method. These defects include scattering particles, pores, fissures, cellular structure and negative crystal. Fig.26 shows some scattering particles in 76%PMN-24%PT crystal. These scattering particles are attributed to incomplete melting of starting materials, whose melting points change from 886°C of PbO to 2830°C of MgO. These particles can be eliminated by pre-synthesizing PMN-PT raw materials. Sometimes, pores occur at the upper part of as-grown boule. They may be related to growth acceleration and can be reduced by decreasing the lowering rate of the crucible gradually during the later stage of crystal growth. The negative crystal with a dimension of about 1mm has been observed in PMN-PT crystal. The negative crystal shows regular crystallographic facets, as shown in Fig.27. The sudden temperature fluctuation sometimes occurs due to the instability of heating elements or ambient temperature, which results in constitutional supercooling and the melt wrapped by grown crystal. The melt droplet then crystallizes gradually and forms negative crystal in the subsequent cooling stage. The negative crystals are attributed to be red PbO and pyrochlore crystals, which are observed by optical microscope [47].

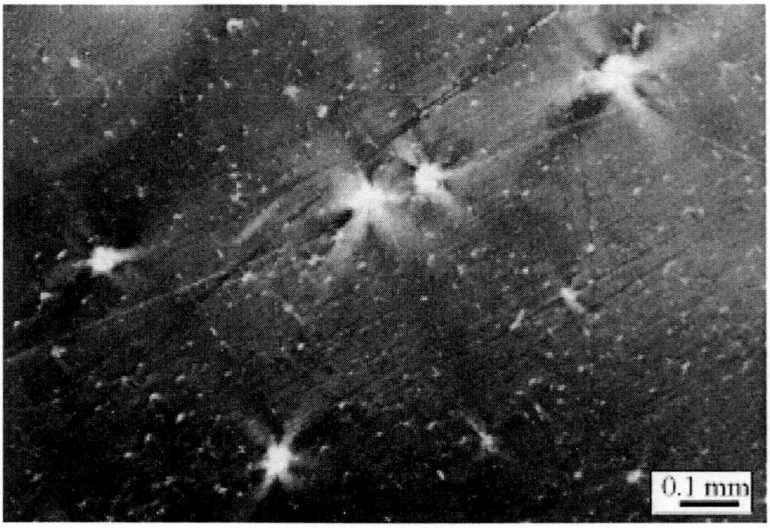


Figure 26. The scattering particles in PMN-PT crystal

Figure 27. The negative crystal in PMN-PT crystal

Compared with the flux method, the vertical Bridgman method shows several advantages in growth of PMN-PT crystals, such as growing large size PMN-PT crystal, restraining the components from vaporizing. Φ2″ PMN-PT crystals have been grown reproducibly by the modified Bridgman method in our institute. But the inhomogeneity of the crystal is still a problem and the crucible leakage makes high cost of the crystal, which seriously hinders its industrial applications.

## 4.3. Growth of Relaxor Ferroelectric PZN-PT Crystals with Pbo Flux

The complete solid solution $(1\text{-}x)Pb(Zn_{1/3}Nb_{2/3})O_3\text{-}xPbTiO_3$ (PZN-PT) is formed in the range $0 \le x \le 0.2$ and the morphotropic phase boundary (MPB) at x=0.09 separates the rhombohedral and tetragonal phases at 25°C [42]. Similar to PMN-PT crystal, PZN-PT has attracted much attention due to its exceptionally good piezoelectric properties at the composition near MPB. The excellent properties render PZN-PT crystals suitable as next-generation electromechanical transducer materials in a broad range of advanced applications, such as actuators, medical ultrasound imaging and underwater communication. Due to its low acoustic impedance [$Z_{33}{}'$ = 22 (106 kg/$m^2$.s)], PZN-PT single crystal seems to be the most promising candidate for array-type medical transducers with a high sensitivity, an improved resolution and a large bandwidth [48].

It is difficult to grow large size and high quality PZN-PT single crystals because of the coexistence of perovskite and pyrochlore phases, the incongruent melting characteristics and multi components of the melt. To grow perovskite PZN-PT crystal, some fluxes have been added to the melts. The crystallization behaviors of 91%PZN-9%PT have been investigated in different composition fluxes, such as $B_2O_3$, $PbF_2$, $BaTiO_3$ and PbO [49]. The results show that $B_2O_3$ and $PbF_2$ fluxes restrain the crystallization of perovskite phases while $BaTiO_3$-PbO composite fluxes are helpful to the crystallization of the perovskite PZN-PT but the crystal yield is low. It is suitable to grow perovskite PZN-PT crystals from the high temperature solution with about 50mol% PbO flux.Many efforts have been made to grow PZN-PT crystals and the flux method has been widely used. However, the conventional flux method has limitations for achieving large single crystals with high quality due to spontaneous nucleation. Kobayashi et al reported that PZN-PT crystal up to 40×40×40mm$^3$ has been grown by a modified flux method with bottom cooling [50]. In recent years, the vertical Bridgman method has been employed and PZN-PT crystals with a dimension of 2 inches in diameter and 1/2 inch in length have been grown from a high temperature solution with a PbO flux [51, 52]. In order to induce a single nucleus, a gas-cooling system is assembled at the bottom of the crucible, as schematically shown in Fig.28. A metal pipe with a sealed top is connected to the bottom of the Pt crucible and oxygen gas flow is induced by a ceramic rod and passed through the metal pipe. When the raw materials are soaked at 1200°C for 6h, oxygen gas flow is suddenly piped in and leads to a rapid crystallization at the bottom of the crucible. Then the lowering mechanism starts to work and the fast-formed nucleus gradually grows into a larger crystal. The surface and the upper part of the ingot are mainly composed of solidified PbO, as shown in Fig.29a. PZN-PT crystals are obtained by leaching the ingot in hot $HNO_3$ for 20h. Fig.29b shows a PZN-PT crystal grown by the modified Bridgman method. The crystal exhibits several crystallographic facets although some sub-crystals have been found at the edge of the boule.

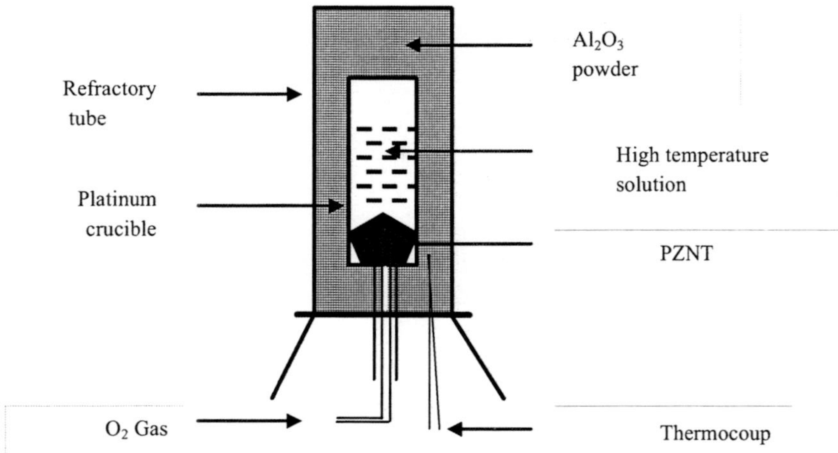


Figure 28. Schematic diagram of the gas-cooling system for Bridgman growth of PZN-PT crystals

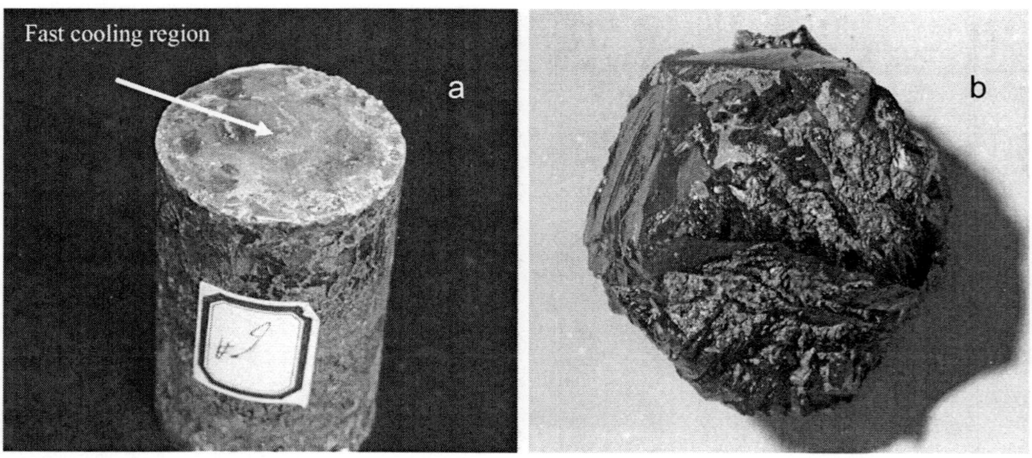


Figure 29. As-grown PZN-PT ingot (a) and crystal (b)

The flux Bridgman method with gas cooling is more suitable to grow large size PZN-PT crystals than the traditional vertical Bridgman method. So far, the maximum PZN-PT crystal of 75mm in diameter, 55mm in height, and 1kg in weight, has been grown successfully by this method [53]. The results seem to indicate that this method has good potential for application in the mass production of PZN-PT single crystals. However, some inclusions of PbO flux and sub-crystals are found in the bottom part and the peripheral regions of the crystal. These inclusions are probably caused by unsuitable growing conditions, such as an inadequate temperature gradient in the furnace and instability of the gas flow. Fig.30 shows some PZN-PT wafers cutting perpendicular to c-axis. Some red particles and inclusions are observed in the wafers. The inclusions in the bottom of the ingot are attributed to the fast crystallization in the initial stage. The parabolic inclusion of about 1mm thickness is formed due to the instability of the gas flow and the delay of solute transport.

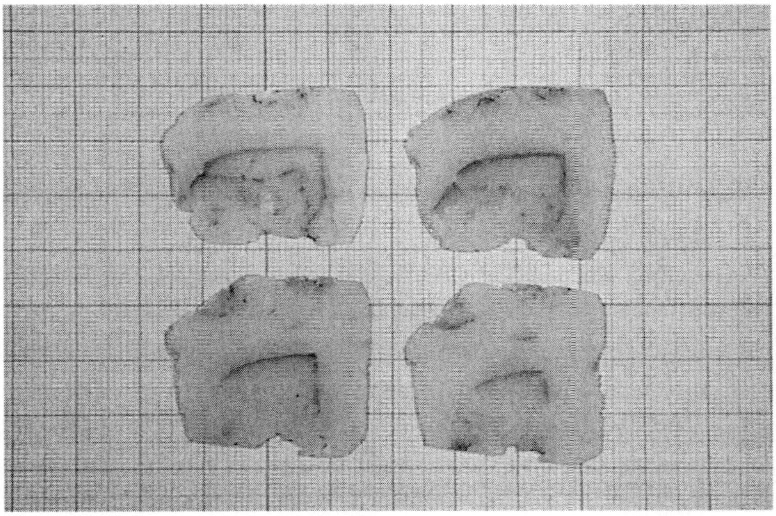

Figure 30. Typical PZN-PT wafers grown by the Bridgman method with bottom cooling

The parabolic inclusion is analyzed by means of Electron Probe X-ray Microscopicanalysis (EPMA). Fig.31 shows the microstructure of the parabolic inclusion. The photo reveals that the inclusion is very complicated and composed of translucent stripes, small particles and red matrix. The compositions of PZN-PT crystal are examined using Philips PW2400 automatic sequential wavelength-dispersive spectrometer and the compositions of the inclusions are obtained by EPMA, as listed in Table 3. The compositions of PZN-PT crystal is very close to the theoretical value of $Pb[(Zn_{1/3}Nb_{2/3})_{0.93}Ti_{0.07}]O_3$ (93%PZN-7%PT). Table 3 confirms that the matrix is mainly composed of PbO (98mol%) while the small particle is composed of 88.11mol% ZnO and 11.89mol% PbO. Thus the particle is suggested to be an accumulation of ZnO, which is the reason why the composition of ZnO in the crystal decreases along growth direction. The translucent region is presumed to be composed of $3PbO{\cdot}Nb_2O_5$, PbO and a small amount of $PbTiO_3$ according to its compositions and microstructures. As mentioned above, the high temperature solution may be surrounded by the crystal in the interface due to sudden temperature fluctuation induced by the instable gas flow. Therefore, ZnO is accumulated to form some particles and $3PbO{\cdot}Nb_2O_5$ crystallizes from the PbO matrix.

**Table 3. Compositions of the inclusions and PZN-PT(mol%)**

| Components | The particle (A) | The translucent stripe (B) | The matrix (C) | PZN-PT crystal (D) |
|---|---|---|---|---|
| PbO | 11.89 | 83.86 | 98.07 | 66.22 |
| ZnO | 88.11 | - | - | 7.31 |
| $Nb_2O_5$ | - | 15.39 | 1.93 | 24.61 |
| $TiO_2$ | - | 0.75 | - | 1.85 |

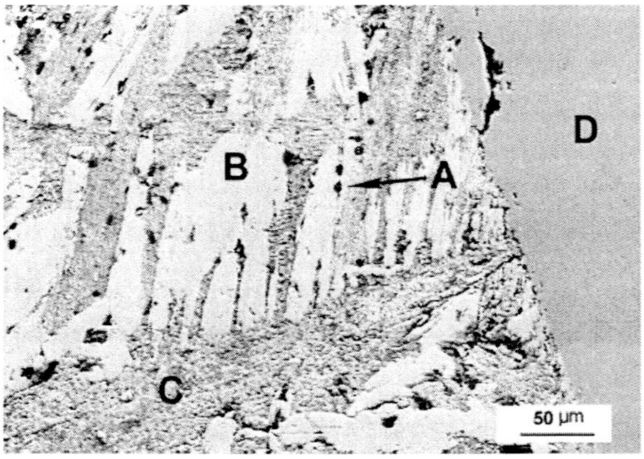


Figure 31. SEM photo of the inclusion region: (A) ZnO particles; (B) translucent region; (C) PbO matrix; (D) PZNT93/7 crystal

Although PZN-PT crystals have been grown in a multi-crucible Bridgman furnace and the yield of PZN-PT crystals has been improved considerably using the gas-cooling system, the inclusions frequently occurred in the crystal influence the quality of PZN-PT wafers seriously and limited its practical application. In order to resolve this problem, a two-step growth process has been developed [54]. In the first stage, the nucleation is induced by the flow of oxygen and a large temperature gradient is formed near the solid-liquid interface. PZN-PT crystals grow fast along the nucleus. After the crystal has been grown about 20mm length, the gas flow is closed and the lowering mechanism stops working for several hours. Accordingly, the previous grown crystal is partially melted and a stable solid-liquid interface is re-established. Then the lowering mechanism continues working and PZN-PT crystal grows again. As-grown ingot is similar to that in Fig.29a, but the morphology of the crystal is very different. Fig.32 shows a PZN-PT crystal grown by the two-step process.

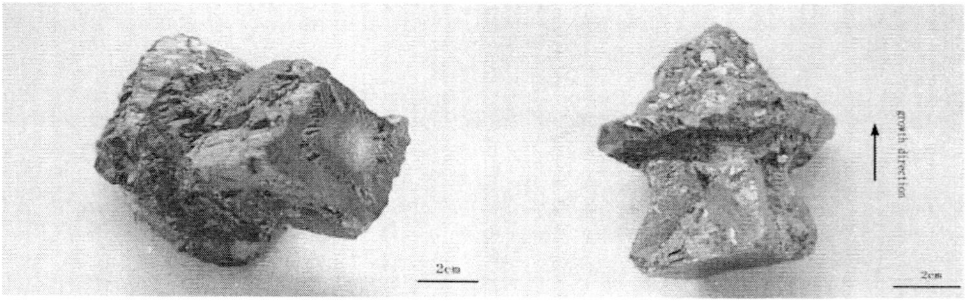


Figure 32. Bottom (left) and side (right) view of PZN-PT crystal grown by the two-step Bridgman method

The ingot of PZN-PT crystal looks like a mushroom and the two growth stages are separated by the morphologies of the different parts. The previous grown crystal is about 38mm × 38mm × 40mm. This means the growth rate of PZN-PT crystal in the first step was much faster than the lowering rate. PZN-PT crystal grown in the second step is about 50mm in diameter and 15mm in height. Because the solute has been exhausted, the PZN-PT crystal grown in the second step is a little shorter. PZN-PT crystal may be grown longer if more raw materials are loaded in the crucible. In this case, the platinum crucible will be eroded seriously with increasing of PbO contents. In fact, the two-step growth process provides a self-seed and a comparatively steady interface to grow high quality PZN-PT crystals. Compared with PZN-PT grown by gas-cooling Bridgman method, the quality of two-step grown PZN-PT has been greatly improved although there are still some small particles in the wafers, as shown in Fig.33. Practical PZN-PT slices up to 25mm ×25mm ×1mm are conveniently obtained by the two-step growth process.

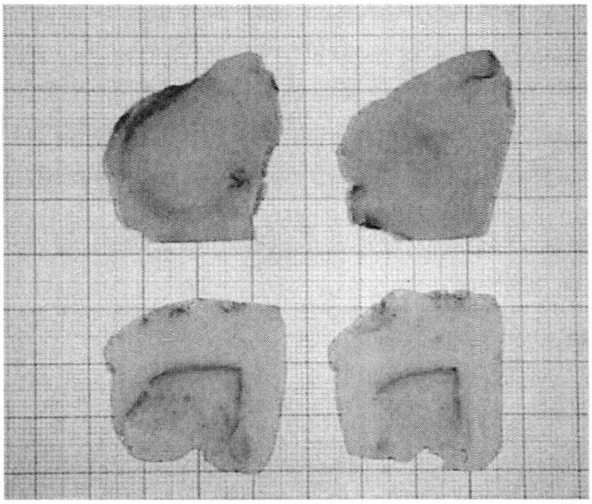

Figure 33. PZN-PT wafers grown by the two-step process

Figure 34. Fissures and pits in the inner wall of the platinum crucible

So far, PbO flux has been proved effectively during Bridgman growth of PZN-PT crystals because of its low viscosity and high solubility. However, it has disadvantages such as toxicity, volatility and a tendency to corrode the platinum crucible. Due to the erosion of PbO on Pt crucible, the leakage of the melt often occurs during the Bridgman growth. Fig.34 shows some fissures and pits in the inner wall of the platinum crucible [55]. The leakage may result in failure of the growth. So the growth temperature is a key parameter for successful growth of PZN-PT crystals.

## 5. Summary

The vertical Bridgman method has been applied to the growth of some oxide single crystals, such as $Li_2B_4O_7$ (LBO), $Sr_3Ga_2Ge_4O_{14}$ (SGG), $LiNbO_3$ (LN) $TeO_2$, $Bi_{12}SiO_{20}$ (BSO), $Bi_{12}GeO_{20}$ (BGO), $Pb_5Ge_3O_{11}$ (PGO), (1-x)$Pb(Mg_{1/3}Nb_{2/3})O_3$-x$PbTiO_3$ (PMN-PT) and (1-x)$Pb(Zn_{1/3}Nb_{2/3})O_3$-x$PbTiO_3$ (PZN-PT), for piezoelectric applications. The multi-crucible Bridgman furnace has been developed for mass production of these crystals and the cost of the crystals can be reduced considerably. The evaporation of the components has been improved by sealing the melts in a limited space of the crucible. In order to grow novel relaxor ferroelectric single crystals, the flux has been led into the vertical Bridgman method and PZN-PT crystal up to 3 inches in diameter has been grown using gas cooling. A two-step grow process has been invented to improve the quality of PZN-PT crystals. Generally, the vertical Bridgman method has following advantages for growing piezoelectric crystals: (1) the crystal shape depends on the shape of the crucible; which means high utility of as-grown crystals during processing standard wafers; (2) the multi-crucible Bridgman furnace is very suitable for mass-production of piezoelectric oxide crystals with lower cost; (3) a small temperature gradient in the solid-liquid interface and in situ annealing process considerably reduce the thermal stress in as-grown crystals; (4) gas cooling at the bottom of the crucible is very helpful to grow large size crystals from high temperature solutions. In one word, the vertical Bridgman method has become an effective technique for industrial growth of piezoelectric oxide crystals.

## Acknowledgement

The authors would like to make acknowledgment to Professor Liao Jingying, Mr. Ge Zengwei, Lu Baoliang, Wu Xianjun, Tong Jian and Xia Zongren for their contributions in crystal growth. The authors also thank Ministry of Science and Technology of China, National Natural Science Foundation of China, Chinese Academy of Sciences and Shanghai Committee of Science and Technology for their financial support in related projects.

## References

[1] Hurtle D.T.J. *Handbook of Crystal Growth*, North-Holland, Amsterdam, 1993, vol.2a, p.53.
[2] Stockbarger D.C. *Rev. Sci. Instrum.* 1963, 7, 133.
[3] Triboulet R. In *Crystal Growth Technology*, Scheel H.J., Fukuda T., Ed., Wiley, England, UK, 2003, pp.373-406.
[4] Jiayue Xu, *J. Synthetic Crystals*, 2002, 31, 298-304.

[5] Gualtieri J.G., Kosinski J.A. and Wilber W.D., *Proceedings of the 1992 IEEE Frequency Control Symposium*, p.724.
[6] Abe H., Saito H. and Ohmura M., *Proceedings of the 1987 IEEE Ultrasonics Ferroelectric and Frequency Control Symposium*, p.91.
[7] Robertson D.S. and Young I.M., *J. Mater. Sci.* 1982, 17, 1729-1735.
[8] Komatsu R. and Uda S., *Materials Research Bulletin*, 1998, 33, 433-440.
[9] Katsumata T. and Ohshima K., *J. Crystal Growth*, 1992, 125, 270-280.
[10] Fan Shiji, Shen Guanshun and Wang Wen, *J. Crystal Growth*, 1990, 99, 811-813.
[11] Jiayue Xu, Shiji Fan, Baoliang Lu, *J. Crystal Growth*, 2004, 264, 260-265.
[12] Ya V. Burak, *J. Crystal Growth*, 1998, 186, 302-304.
[13] Xu Jiayue, Xu Yibin, Fan Shiji and Wang Wen, *J. Inorganic Materials*, 1993, 8, 417-420.
[14] Whatmore R.W., Shorrocks N.M., et al, *Electronic Letters*, 1981, 17, 11-13.
[15] Bhalla A.S. et al, Jpn. *J. Appl. Phys.*, 1985, 24(Suppl.), 727.
[16] Shiosaki T., Adachi M., Kobayashi H., Araki K. and Kawabata A., *Jpn J. Appl. Phys.*, 1985, 24(suppl), 25-27.
[17] Abe H., Ohmura M. and Saitou H., *Proceedings of the 1994 IEEE international Frequency Control Symposium*, p.289.
[18] Jiayue Xu, Baoliang Lu, Shiji Fan, *Proceedings of t*he 7th International Conference on Solid-State and Integrated-Circuits Technology, 2004, Beijing, China, p.2198.
[19] Gotalskaya A.N., Drezin D.I., Bezdelkin V.V., Stassevich V.N., *Proceedings of the 1993 IEEE Frequency Control Symposium*, p.339.
[20] Buzanov O.A., Naumov A.V., Nechaev V.V., Knyazev S.N., *Proceedings of the 1996 IEEE Frequency Control Symposium*, p.131.
[21] Wu An-Hua, Xu Jia-Yue, Zhou Juan, Fan Shiji, *J. Chin. Ceram. Soc.*, 2003, 31, 480-484.
[22] Shimamura K., Takeda H., Fukuda T., *J. Cryst. Growth*, 1996, 163, 388-392.
[23] Juan Zhou, Jiayue Xu, Wangxiang Hua, Shiji Fan, *Materials Sciences and Engineering B*, 2004, 106, 213-217.
[24] Ballman A.A., *J. Am. Cerm. Soc.*, 1965, 48, 112.
[25] Hongbing Chen, Haiping Xia, Jinhao Wang, Jianli Zhang, Jiayue Xu and Shiji Fan, *J. Crystal Growth*, 2003, 256, 219-222.
[26] Standifer E.M., Jundt D.H., Norwood R.G., Bordui P.F., *Proceedings of the Annual IEEE International Frequency Control Symposium*, Pasadena, USA, 1998, p.470.
[27] Grewal P.K. and Lea M.J., *J. Phys. C: Solid State Phys.* 1983, 16, 247.
[28] Xu Xuewu, *Ph D dissertation*, Shanghai Institute of Ceramics, Chinese Academy of Sciences, Shanghai, 1991.
[29] Xu Xuewu, Liao Jingying, Shen Bingfu, Shao Peifa, Chen Xianqiu and He Chongfan, *J. Crystal Growth*, 1993, 133, 267-272.
[30] Xu Xuewu, Liao Jingying, Shen Bingfu, Shao Peifa, Chen Xianqiu and He Chongfan, *J. Chin Ceram. Soc.*, 1994, 22, 579-585.
[31] Tanguay A.R., Mroczkowski Jr. S. and Barker R.C., *J. Crystal Growth*, 1977, 42, 431-434.
[32] Aiqiong Zhang, Jiayue Xu and Shiji Fan, *Chemical Research*, 2004, 15, 1-5.
[33] Veber P., Mangin J., Strimer P., Delarue P., Josse C., Saviot L., *J. Crystal Growth*, 2004, 270, 77-84.

[34] Arlt G., Schweppe H., *Solid. State Commun.* 1968, 6, 783-784.
[35] Ledbetter H., Leisure R.G., Migliori A., Betts J. and Ogi H., *J. Appl. Phys.*, 2004, 96, 6201-.
[36] Grabmaier J.G., Plattner R.D., Schieber M., *J. Crystal Growth*, 1973, 20, 82-88.
[37] Chu Yueqing, Ge Zengwei, Wu Guoqing, Yin Xueji and Tang Linyao, *J. Synthetic Crystals*, 2004, 33, 810-812.
[38] Xianjun Wu, Jiayue Xu, Jingzhong Xiao, Anhua Wu, Weiqing Jin, *J. Crystal Growth*, 2004, 263, 208-213.
[39] Sugii K, Iwasaki H. and Miyazawa S., *Mat. Res. Bull.* 1971, 6, 503-511.
[40] Yamashita Y., Hosono Y., Harada K. and Ichinose N., *Jpn. J. Appl. Phys.* 2000, 39, 5593-5596.
[41] Kuwata J., Uchino K. and Nomura S., *Ferroelectrics*. 1981, 37, 579-582.
[42] Kuwata J., Uchino K. and Nomura S., *Jpn. J. Appl. Phys.* 1982, 21, 1298-.
[43] R.F. Service, *Science*. 1997, 275, 275.
[44] Haosu Luo, Guisheng Xu, Pingchu Wang, Zhiwen Yin, *Ferroelectrics*, 1999, 231, 97-102.
[45] Haosu Luo, Guisheng Xu, Haiqing Xu, Pingchu Wang and Zhiwen Yin, *Jpn. J. Appl. Phys.* 2000, 39 (9B), 5581-5585.
[46] Zawilski K.T., Custodio M.C.C., DeMattei R.C., Lee S.G., Monteiro R.G., Odagawa H., Feigelson R.S., *J. Crystal Growth*, 2003, 258, 353-367.
[47] Guisheng Xu, Haosu Luo, Haiqing Xu, Zhenyi Qi, Pinchu Wang, Weizhuo Zhong, Zhiwen Yin, *J. Crystal. Growth*, 2001, 222, 202-208.
[48] S. Saitoh, T. Takeuchi, T. Kobayashi, K. Harada, S. Shimanuki and Y. Yamashita: *Jpn. J. Appl. Phys.* 1999, 38(5B), 3380-3384.
[49] Jiayue Xu, Shiji Fan, Renying Sun, *J. Chin. Ceram. Soc.*, 2002, 30, 722-725.
[50] Kobayashi T, Shimanuki S, Saitoh S, Yamashita Y, *Jpn. J. Appl. Phys.* 1997, 36(9B), 6035-6038.
[51] K. Harada, Y. Hosono, Y. Yamashita, K. Miwa, *J. Crystal Growth*, 2001, 229, 294-298.
[52] Jiayue Xu, Jian Tong, Minli Shi, Anhua Wu and Shiji Fan, *J. Crystal Growth*, 2003, 253, 274-279.
[53] M. Matsushita, Y. Tachi, K. Echizenya, *J. Crystal Growth*, 2002, 853, 237-239.
[54] Jiayue Xu, Xianjun Wu, Jian Tong, Minli Shi, Guoxing Qian, submitted to *J. Crystal Growth*.
[55] Jiayue Xu, Shiji Fan, Baoliang Lu, Jian Tong, Aiqiong Zhang, *Jpn. J. Appl. Phys.*, 2002, 41, 7000-7002.

In: Trends in Crystal Growth Research
Editor: George V. Karas, pp. 133-159
ISBN 1-59454-541-3


*Chapter 6*

# PLASMON AND CORE EXCITATIONS OF CARBON ALLOTROPES

***Hansoo Kim****

Department of Materials Science and Engineering, University of Pennsylvania, Philadelphia, PA 19104

## ABSTRACT

Carbon forms various allotropes from amorphous carbon to recently-discovered fullerenes and carbon nanotubes. In spite of the elemental identity, carbon allotropes show different, albeit similar in many respects, physical properties as a result of their structural dissimilarities. Electron energy loss spectroscopy has been one of the most important tools used for measurements of their electronic properties because it can show the elementary excitations of solids in a wide energy range with a momentum transfer or with a high spatial resolution. In this study, previous reports on plasmon and core excitations of carbon allotropes investigated by electron energy loss spectroscopy are reviewed. Therefore, the positions of their $\pi$ , $\sigma$, $\pi + \sigma$ plasmons, and intraband transitions along with carbon 1s core level absorptions are compared and the reasons for similarities and differences in these excitations and for conflicts between previous reports are discussed. Furthermore, dependence of the plasmons on momentum transfer and spatial locations are reviewed as well as dielectric functions and joint energy density of states of the carbon materials. Also, for comparison electron energy loss spectra in multiwalled carbon nanotubes and amorphous carbon were recorded and assessed. Collections of spectra were recorded by a Gatan PEELS spectrometer attached to transmission electron microscopy by scanning an electron probe across carbon nanotubes or positioning it on an amorphous carbon film.

Carbon has six electrons with two electrons in the K shell and the other four electrons in the L shell. The K shell consists of the 1s orbital close to the nucleus while the L shell is composed of the 2s and the 2p orbitals. Therefore, the carbon atom has ground state electronic configuration $1s^2\ 2s^2\ 2p^2$. Electrons in the 2s and the 2p orbitals can hybridize and

* E-mail: hansoo@seas.upenn.edu, hsluminesc@hotmail.com ; Phone: 215-573-1179 ;FAX:215-573-2128

form $sp^1$, $sp^2$, and $sp^3$ bonds and these hybrid orbitals are directional in one, two, and three dimensions, respectively. In nature carbon atoms can arrange themselves into various allotropes which are amorphous carbon, graphite, diamond, fullerene, carbon nanofoam, lonsdaleite, and carbon nanotubes[1]. Among these carbon materials, graphite has $sp^2$ bonding corresponding to the hybridization of the 2s orbitals with $2p_x$ and $2p_y$ orbitals into a structure of trigonal and planar σ bonds. The other $2p_z$ orbital in the carbon atom forms a π bond, extending above and below the σ bond plane. Therefore, in the layered crystal structure of graphite, each layer consists of the strong σ bonds and the weak π bonds are responsible for the interlayer bonding. This weak π bond of graphite determines many physical properties of the material[2]. On the other hand, diamond is formed based upon $sp^3$ bonding for which the 2s orbitals hybridize with three $2p_x$, $2p_y$, and $2p_z$ orbitals to compose tetragonal σ bonds. Since all 2p orbitals participate in the hybridization, no π bond is derived when carbon atoms are assembled into diamond. In contrast to the carbon materials based on pure $sp^2$ and $sp^3$ bonding, amorphous carbon is possessed of both $sp^2$ and $sp^3$ bonding while carbon nanotubes have curvature-modified $sp^2$ bonding whose deviation from $sp^2$ bonding is dependent on their diameter. $C_{60}$ fullerenes also have curvature-modified $sp^2$ bonding though in the case of $C_{60}$ fullerenes the size is invariable. This curvature-modified $sp^2$ bonding takes on a bonding state of $sp^{2+x}$, so that, for instance, $C_{60}$ fullerenes are regarded as composed of hybridized $sp^{2.275}$ bonding[3]. The differences in the electronic and crystallographic configurations of these carbon materials bring about variations in their physical properties. For example, diamond is transparent to visible light and very hard equally in all crystallographic directions while graphite is black and slippery in one direction. The π electrons forming the π bonds contribute electric and thermal conductivity of graphitic materials such as graphite, fullerenes, and nanotubes. Therefore, graphite is semimetal, face centered cubic $C_{60}$ fullerenes are semiconducting with a direct band gap, and single wall nanotubes can be semiconductor or metal[4-7]. The π and σ electrons in carbon materials can behave as delocalized and collective electron gases or plasmons, whose response to the electron beam or the optical light gives a lot of information about electronic and optical properties of the materials such as dielectric functions and joint energy density of states. The electrons can also excite and deexcite themselves individually by the external stimuli, changing their electronic energy states depending upon the quanta of the external energy and selection rules. These types of transitions between energy states are called interband transitions. The interband transitions are considered to induce the plasmonic transitions for the bulk as well as for the surface[8-10]. In fact, the π plasmon peaks in various carbon materials stem from the π - π* interband transitions and the π + σ plasmon peaks originate primarily from the σ - σ* interband transitions. Still another kind of transitions of the electrons can be seen in carbon materials, which are called core transitions. These transitions occur between the 1s core levels and the unoccupied antibonding states σ* and π* of a carbon atom. From core transitions, information about electronic and chemical states of carbon atoms can be obtained.

To find out the responses of carbon materials to the external stimuli, diverse analytical tools have been employed. Therefore, scanning tunneling microscopy using electric field, x-ray absorption fine structure, Fourier transform infrared, elipsometry, and Raman scattering using x-ray, infrared and visible light have been exploited[11-16]. Its versatility and wide energy range of the source electron beam made electron energy loss spectroscopy (EELS) one of the most frequently-used measures[17,18]. In this study we focus on the plasmon and core excitations of carbon materials obtained by EELS even though in some cases the results will be compared with those taken by other instruments and those calculated from simulations. EELS has two operational modes, one of which is a reflection mode and the other is a transmission mode. The transmission mode has been the primary mode used for the acquisition of the plasmon and core excitations of carbon materials. However, the reflection mode has been also used to get EEL spectra of fullerenes.

Two types of EELS have been employed for obtaining electron energy loss (EEL) spectra of carbon materials. The first one may be named a dedicated EELS (or high resolution EELS - HREELS) which is only used for acquisition of EEL spectra. With this type of EELS, the energy resolution (the full width at half maximum of a zero loss peak) is adjustable and excellent (down to even less than 1 meV). Therefore, even vibrational or phonon modes of a specimen can be measured by this dedicated EELS or HREELS. Also, momentum transfer can be used as another variable along with energy. Momentum transfer is a difference in the momenta of incident and scattered electrons. EEL spectra acquired from a material with a varying momentum transfer and energy can constitute a complete set of data for understanding the elementary excitations of the material[19,20]. However, this type of EELS is not capable of probing individual nanometer-sized materials such as fullerenes and carbon nanotubes, and acquiring their location-specific response. Namely, it shows a poor spatial resolution. The other type of EELS is attached to transmission electron microscopy (TEM) or scanning transmission electron microscopy (STEM). Compared to the first one, this type of EELS gives relatively poor energy resolution (higher than 0.1 eV: ~ 0.3 – 2 eV depending on the electron source). Momentum transfer can be also used as another variable but generally the momentum resolution of electron-microscope-electron-spectroscopy system is not as good as that of dedicated EELS. However, its spatial resolution is normally down to sub-nanometer scale, so that location-specific response of nanometer size specimen such as fullerenes and carbon nanotubes can be obtained. Moreover, another beneficial merit is that each nanometer-sized object can be observed by TEM or STEM while EEL spectra are being captured.

In this paper plasmon and core excitations of carbon allotropes (amorphous carbon, graphite, diamond, fullerenes, and nanotubes) taken by both types of transmission EELS are reviewed. From these data, how electrical and optical properties can be derived from EEL spectra is also reviewed. In addition, EEL spectra from purified multiwalled carbon nanotubes (MWNTs) and amorphous carbon were empirically obtained using a commercial STEM-attached EELS.

MWNTs used in this study were synthesized at 700 °C by chemical vapor deposition using iron as a catalyst and cooled to room temperature at the rate of about – 20 °C/min. After ~ 6 mol% of ferrocene ($Fe(C_5H_5)_2$) was dissolved in xylene ($C_6H_4(CH_3)_2$), which were used as a catalyst and a carbon sources, the solution was preheated at ~ 200 °C and carried into the reactor by an Ar/10% $H_2$ with the rate of 6 l/min (STP) during synthesis. CNTs grown perpendicular to a quartz plate in the form of thick mats were collected. These catalytically grown CNTs contain a rod shape of iron particles inside the cavities[21]. Some CNTs were annealed at 2000 °C for 40 min in an Ar atmosphere to remove iron particles and to graphitize MWNTs[22]. EEL spectra were collected from a few spatial points of a purified MWNT along its diameter. Also, an amorphous carbon film on a holy carbon support film for a specimen grid was used as an amorphous carbon sample. TEM (JEOL 2010F) equipped with Gatan parallel electron energy loss spectroscopy was employed to record the response of purified CNTs to the electron beam. EEL spectra were collected in a STEM mode using a 20-

μm condenser aperture and a 0.5-nm electron beam probe. For low and core loss spectroscopy, 0.6-mm and 2-mm collection apertures were used and the energy resolution for each aperture was about 1.5 and 2 eV, respectively.

## AMORPHOUS CARBON

Figure 1 shows a TEM micrograph, and plasmon and core excitations of amorphous carbon. In the inset of Fig. 1 (a) the typical phase contrast structures of a thin amorphous carbon film or a weak phase object are shown, whose size is dependent on the wavelength (or the operating voltage) of the electron beam and the defocus value. The low loss spectrum was obtained after removing the zero loss peak and plural scattering through the Fourier-Log deconvolution. This low loss spectrum displays a similar trend with the energy loss function, that is, the inverse imaginary part of the complex dielectric function [$\mathrm{Im}(-1/\varepsilon(q, E))$, q: transferred momentum, E: energy] when the collection angle is larger than the convergence angle. The collection angle is set by the size of the entrance aperture (or collection aperture) of the EELS system and the convergence angle is determined by the angle between the z-axis of the microscope column and the converging electron beam[23]. From the low loss spectrum deconvoluted by the Fourier-Log method and transformed by the Kramers-Kronig method, the imaginary part of the dielectric function can be deduced, by which the absorption of energy or dielectric loss by the specimen can be also estimated[24]. The peak at about 1.7 eV in Fig. 1 (a) is an artifact of the deconvolution. This will be further explained in detail below. Other peaks in the figure are shown at about 6 and 23 eV which correspond to the $\pi$ and $\pi + \sigma$ plasmons, respectively. In fig. 1 (b) the pre-edge peak at about 285 eV is characteristic of the excitation from the 1s core levels to vacant $\pi^*$ states in the $sp^2$ – bonded carbon atoms. The peak at about 292 eV indicates from the 1s core levels to vacant $\sigma^*$ states in the $sp^2$ and $sp^3$ – bonded carbon atoms. Typical features in the core excitations of the amorphous carbon are that the gap between the $\pi^*$ and $\sigma^*$ transitions is not wide and above 295 eV no specific fine peaks are further observed.

As mentioned above, amorphous carbon is mainly composed of $sp^2$ and $sp^3$ bonding although for low density carbon films with a relatively small portion (less than 10%) of $sp^3$ bonding, sp bonding is also detectable[25]. Since the ratio of $sp^2$ to $sp^3$ bonding in amorphous carbon films plays a crucial role in their physical (especially mechanical) properties, methods have been devised, by which the ratio is evaluated using electron energy loss near-edge fine structure (ELNES) of the carbon K-edge[26-29]. According to the methods[27,28], besides the $\pi^*$ and $\sigma^*$ transitions at 285 and 292 eV, another peak attributed to a molecular transition at 287 eV is considered. The molecular transition[30-32] has a varying intensity depending on the intramolecular bond length and the change in the molecular-like property of a sample. Using three Gaussian and/or Lorentzian functions the $\pi^*$ and $\sigma^*$ peaks and the peak at 287 eV of the experimental core EEL spectrum are fitted to reach the highest value for the likelihood. With the integrated areas of the Gaussian and/or Lorentzian functions and a reference (graphite at the magic angle or $C_{60}$ fullerite crystals) the ratio of $sp^2$ to $sp^3$ bonding in amorphous carbon films can be calculated.

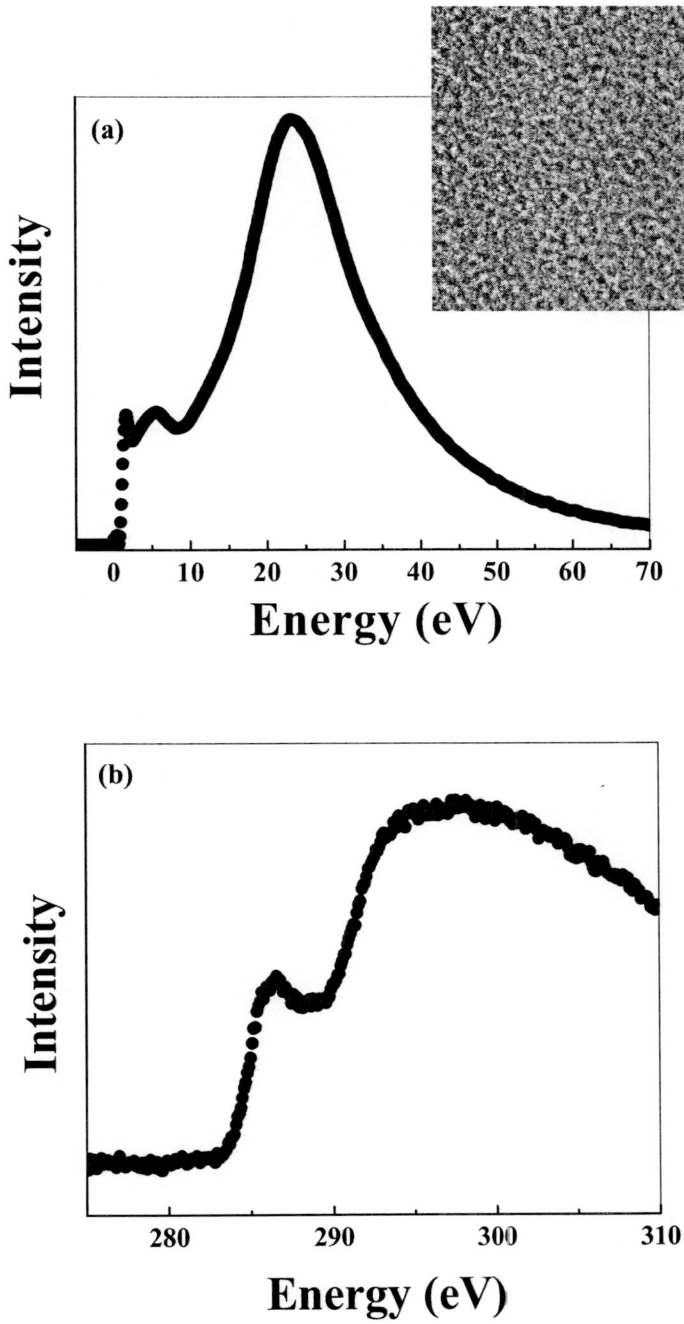


Figure 1. (a) Valence and (b) Core EEL spectra of amorphous carbon. The inset of (a) is a TEM micrograph of amorphous carbon.

Rafferty and Brown developed a theory by which it was proposed that from the energy loss function or the low loss spectrum it can be determined whether the sample has a direct band gap, an indirect band gap, or both[33]. According to the authors, $E \cdot Im(\varepsilon) = k \cdot M^2 \cdot \rho$ where $Im(\varepsilon)$ is the imaginary part of the dielectric function, E is energy loss, k is a proportional constant, M is the matrix element, and ρ is the joint density of state (JDOS). JDOS is the convolution of the valence and conduction energy densities of states. JDOS is proportional to $(E - E_g)^{1/2}$ for direct and indirect band gap transitions, where $E_g$ is energy band gap. According to the authors, $M^2$ is also proportional to $(E - E_g)$ for indirect band transitions while it is a constant for direct band transitions. Therefore, the $E \cdot Im(\varepsilon)$ term is proportional to $(E - E_g)^{1/2}$ for direct band gap transitions and to $(E - E_g)^{3/2}$ for indirect bang gap transitions. Thus, with $Im(\varepsilon)$ derived out of single inelastic scattering data, $E \cdot Im(\varepsilon)$ can be simply calculated to show the trend for the product of the square of the matrix element and the JDOS. Along with the fact that the onset of energy loss function is an indicator of the energy band gap of the sample[18], this may be a very informative method to understand its electronic properties. There in the reference[33] a few materials were tested by this method and compared with previous results, showing a very close match though for the extraction of single scattering data Fourier-ratio deconvolution method was employed instead of Fourier-Log deconvolution method. This method was also applied to amorphous carbon, tetrahedral amorphous carbon (or diamond-like carbon: DLC), and $C_{60}$ fullerite[34]. The results show that tetrahedral amorphous carbon and $C_{60}$ fullerite behave like semiconductors with direct band gaps whose values are 2.63 and 1.59 eV, respectively. On the other hand, in the case of amorphous carbon the way it was measured hampered the authors from obtaining information for its electronic structures. With an STEM-attached EELS it is generally difficult to extract reliable data in the vicinity of zero loss energy. Therefore, from a material with a small band gap like amorphous carbon it is hard to know the band gap energy and the property of the band gap using the STEM-attached EELS. This difficulty and its derivative problems will be further discussed below when EEL spectra from multiwalled carbon nanotubes are dealt with.

## Graphite

Graphite is one of the materials investigated most extensively by many researchers for a long time by various analyzing tools[35-43]. In graphite hexagonally-arrayed carbon atoms form layers which stack in an ABA sequence with the c/a ratio of about 2.73[44]. Often, graphite in an ABC stacking sequence with a rhombohedral symmetry is also found[45]. One of the most conspicuous features in graphite is its anisotropic array of carbon atoms. As can be easily understood in Fig. 2 (a), graphite can look different depending on the viewing directions. Accordingly, it is necessary to obtain the excitations of graphite at various angles to understand its complete response to the external stimuli. Marinopoulos et al obtained valence EEL spectra using a dedicated high resolution EELS, and the zero loss peak and contributions from plural scattering events were removed to obtain energy loss functions for comparison with their simulation results[43]. The energy loss functions in Fig. 2 (b) show the angular dependence of graphite in detail. The π plasmons are seen at ~ 7 eV and the π + σ plasmons are at ~ 27 eV when the c-axis of graphite and the momentum transfer between incident and scattered electrons are perpendicular to each other. In the figure, when the c-axis is parallel to the momentum transfer, the energy loss function (the bottom panel) is a function calculated using the crystal local field effects within the random phase approximation while the other ones are all obtained from empirical EELS measurements. As the angle between the

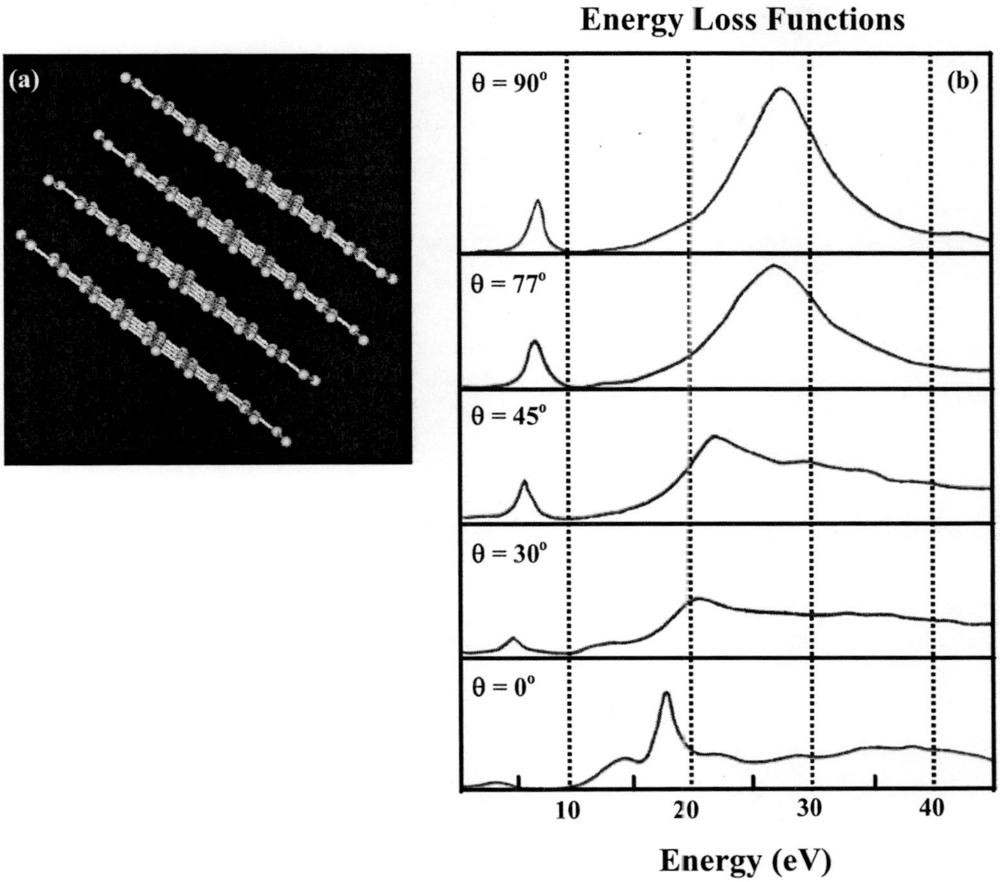


Figure 2. (a) An image of graphite (from www.edinformatics.com/math_science/carbon.htm). (b) Energy loss functions of graphite in various geometries. θ is the angle between the c-axis of graphite and the momentum transfer.

c-axis and the momentum transfer changes, the intensities and positions of the π and π + σ plasmon peaks are modified noticeably. Therefore, as the angle between the c-axis and the momentum transfer decreases from the top to the bottom panels, the π plasmon excites less strongly and the positions shifts to lower energies. The π + σ plasmon also decays fast showing a redshift while in the extended energy range of ~ 18 to 45 eV a broad excitation appears. These modifications in energy loss functions with the orientation of graphite are attributed to the anisotropy of the dielectric functions (Fig. 2 (c)). Namely, as the momentum transfer aligns parallel to the c-axis of graphite, the oscillator strength related with the π - π* and σ - σ* interband transitions is greatly suppressed. As mentioned above, the two interband transitions are deeply involved in the π and π + σ plasmons, respectively. This explains the decay of the plasmons. Instead, at this orientation oscillators responsible for the π -σ* and σ - π* interband transitions resonate strongly to the electron beam at ~ 11 and 16 eV. Also, the graphene planes respond to the electron beam more like isolated individual planes in this geometry, so that the energy loss function is very similar to the imaginary part of the dielectric function. Therefore, we can see peaks in the energy loss function in the range of 10 – 17 eV. The broad peak in the range of ~ 18 to 45 eV is considered to be a continuum of electron-hole excitations[43]. Figure 2 (d) presents core EEL spectra of graphite[46]. The π* and the lowest σ* peaks in these spectra are seen at almost the same positions with amorphous carbon (~ 285 and ~ 291 eV, respectively). It was found that the threshold of the σ* transition is caused by an exciton and a considerable part of the π* transition is also excitonic[40]. The differences between amorphous carbon and graphite are that for graphite the peaks of the π* and σ* transitions are clearly separated to see the gap between the two transitions, and the post-threshold features are unambiguous to show three additional σ* peaks. However, the overall peak intensities and shapes are affected by the orientation of graphite and experimental conditions (convergence and collection angles). In the Fig. 2 (d) the difference of core EEL spectra by the experimental conditions is shown[46]. The relative intensities of the 1s to π* and σ* transitions by the electron beam, or the differential scattering cross-sections for the excitations, are supposed to be also relevant to the angle between the orientation of graphite and the incident electron beam. Rosenberg *et al*[47] measured carbon K edge photo-absorption spectra of single crystal graphite with the angle. According to their results[47] the spectra successfully follow dipole selection rule, that is, the carbon 1s to σ transition is allowed when the polarization vector is normal to the c-axis of graphite while the carbon 1s to π transition is excited when the polarization vector is parallel to the c-axis of graphite. In Fig. 2 (d) the spectra obtained with the c-axis of graphite parallel to the TEM column also show the same trend. The solid line corresponds to the experimental conditions of 0 and 0.72 mrad convergence and collection angles, respectively while the dotted line to conditions of 7.4 and 2.6 mrad convergence and collection angles, respectively. Therefore, when the convergence and collection angles are small (or the parallel weighing fraction is close to 1) and thus when almost all the electrons into and out of graphene layers are aligned parallel to the c-axis of graphite, the π* transitions are greatly enhanced (solid line). On the other hand, when the component of electrons perpendicular to the c-axis is increased and thus the parallel weighing fraction is close to zero (or the convergence angle is large) the σ* transitions is also intensive (the dotted line). However, it was found that for a given convergence angle a so-called magic angle can be determined at a specific collection angle at which the relative intensities of the π* and σ* transitions are not affected by the orientation of graphite with regard to the incident electrons[46]. Therefore, for some information such as the dichroism in EEL spectra[48], specific experimental conditions with which the anisotropy of core excitations of graphite is maximized can be used while for another category of information on the $sp^2 - sp^3$ ratio in amorphous carbon the magic angle can be exploited.

From inelastic single-scattering data extracted out of a valence EEL spectrum, Im(-1/ε) can be calculated with the knowledge of the zero loss intensity, the specimen thickness, and the collection angle, the velocity of incident electrons, energy loss, and a few other physical constants[18]. This is the reason that the deconvoluted low loss spectrum shows the similar trend with the energy loss function, that is to say, they have a proportional relationship with each other. Furthermore, using the Kramers-Kronig transformation of Im(-1/ε), its real part (Re(1/ε)) can be also calculated after angular corrections and corrections for incident beam convergence. Additionally, other optical properties such as absorption coefficient and reflectivity can be obtained by the Kramers-Kronig analysis. These imaginary and real parts of the dielectric function give a few important clues to understanding the reaction of a

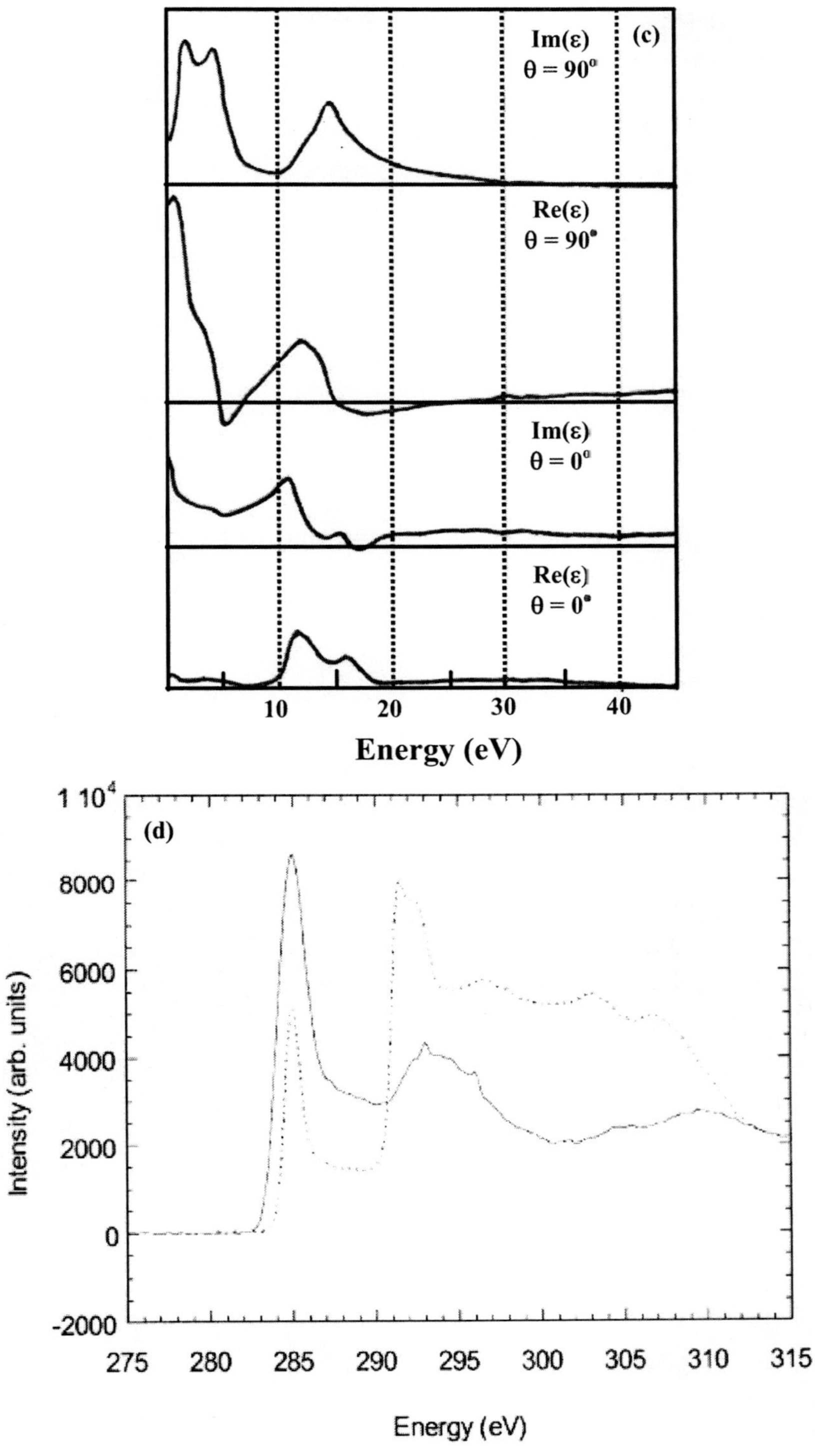


Figure 2. (c) Imaginary and real parts of dielectric functions of graphite. (d) Core loss spectra of graphite taken with different aperture systems (from reference 46) (solid line: convergence angle – 0 mrad, collection angle – 0.72 mrad, dotted line: convergence angle – 7.4 mrad, collection angle – 2.6 mrad).

material with the incident electrons and its electronic structures. First of all, Im(ε) of the dielectric function shows the interband transitions of a material as does the optical absorption. Therefore, in Fig. 2 (c) the peaks in Im(ε) correspond to interband transitions of graphite. The Im(ε) graph in the upper panel, corresponding to the in-plane excitations, show broad excitations in the range of 0 to ~ 7 eV with two peaks at ~ 2 and ~ 4 eV. These excitations are caused by the π - π* interband transitions. On the other hand, the excitation at about 14 eV is brought about by the σ - σ* interband transitions. Another information we can acquire from the dielectric function is that the energies of plasma oscillations can be determined where Re(ε) is equal to zero and Im(ε) is small. In the two upper panels of Fig. 2 (c), the Re(ε) graph shows two zeros at 5 and 7 eV while the oscillator strength in the Im(ε) graph is severely decreased above 4 eV. Therefore, a plasmon oscillation is expected to appear at 7 eV in the valence EEL spectrum of graphite, which is in fact shown as the π plasmon peak in Fig 2 (b). In the same manner another plasmon oscillation can be also anticipated at 23 eV corresponding to the π + σ plasmon in the valence EEL spectrum. There is still another information that can be deduced from the Re(ε) and Im(ε) pair, which is the shape of a plasmon peak. Namely, at the energies where Re(ε) is zero the width of a plasmon peak is proportional to the value of Im(ε) and the height of a plasmon peak has a linear relationship with the inverse of Im(ε)[49]. As mentioned above, this is because Im(ε) is involved in dielectric loss and especially at the energies where Re(ε) is zero it is related with plasmon damping due to the electron interband transitions[20]. The damping gives rise to an increase in the width of the plasmon peak, possibly shifting the peak position in the valence EEL spectrum from the energies corresponding to the zeros of Re(ε).

According to the *ab initio* calculations of Marinopoulos *et al*, as the value of the momentum transfer is increased along directions parallel to the graphitic basal plane, the π and π + σ plasmons decay with the peak positions shifting to higher energies[50]. Also, for large values of the momentum transfer the zeros of Re(ε) do not match with the plasmon energies because of the increase in the value of Im(ε) broadening the π + σ plasmon peak. The authors explain that the π + σ plasmon disperses to higher energies because the crystalline local field effects of graphite shift the oscillator strength to higher energies in the Im(ε) graph, changing the peak shape and yielding additional damping.

## DIAMOND

Diamond is characteristic of the saturated carbon bonding, so that each carbon atom has four nearest carbon neighbors forming a tetrahedron. In diamond no π bond is found, so that the π and π + σ plasmons are not seen in the low loss spectrum. In Fig. 3 (a) the atomic arrangement of carbon atoms in diamond is seen along with the tetragonal bonding of each atom. In spite of the absence of the excitations related with the π plasmon, relatively various peaks have been reported from the valence EEL spectrum. Figure 3 (b) shows a representative valence EEL spectrum collected from diamond[51]. Electronic excitations were identified at about 6.5, 12.5, 16.5, 22.7, 34, and 63 eV. For clear vision of the first two peaks, Fig. 3 (c) was presented where the graph shows Im(ε) obtained from optical reflectance measurements[52]. According to various reports[51-56], peaks at 6.5, 12.5, and 16.5 eV are assigned to interband transitions. These interband transitions may be confirmed in Fig. 3 (c) with the peak at 16.5 eV embedded in the peak at 12.5 eV and making the latter peak asymmetric. About the interpretation of the peak at 22.7 eV there has been some controversy, so that Roberts and Walker[53] assigned it to an interband transition while Lurie and Wilson[55] attributed it to both an interband transition and surface plasmons. On the other hand, Maguire[56] asserted that the peak should be assigned to surface plasmons because the interband transition ($\Gamma'_{25} - \Gamma_1$) mentioned above is not an allowed transition from symmetry considerations. This discrepancy was explained by the fact that even though the dipole transition is forbidden, its quadruple transition is allowed and the other fact that there are energy states close to $\Gamma'_{25}$ and $\Gamma_1$, respectively and a few dipole transitions between these neighboring energy states with a energy gap similar to $\Gamma'_{25} - \Gamma_1$ are allowed[51,54]. Therefore, the peak at 22.7 eV can be primarily assigned to the interband transition. The peak at 34 eV originates from a bulk plasmon and the one at 63 eV from two bulk plasmons[55,56]. However, Waidmann *et al*[57] considered the peak at 34 eV to be a complex system of a plasmon excitation and interband transitions although it takes on a more collective (or plasmonic) character. Also, they showed the electronic excitations of diamond with varying momentum transfers[57]. Even though a material with a cubic crystal structure shows isotropic response in optical spectroscopy or to the electron beam with low values of the momentum transfer, the excitations of diamond may become anisotropic depending upon the crystallographic directions especially at a high momentum transfer since electrons from carbon form a greatly inhomogeneous and localized arrangement. According to their results[57], in fact, the response of diamond to the electron beam shows different behavior with respect to the crystallographic directions especially for high momentum transfers. When the momentum transfer increases its value along the <100> crystal direction, the peak at 34 eV shows a relatively large dispersion compared to the case that it increases along the <111> and <110> crystal directions.

Figure 3 (d) presents the carbon K-edge spectrum of diamond measured by EELS with a high energy resolution (0.19 eV)[58]. In the figure, though partially cut off, the core excitation from carbon 1s to $\pi^*$ energy states, which is seen in the range of about 283 to 287 eV for graphite, is not observed as a result of the saturated carbon bonding of diamond. As is the case with graphite (see above), the initial peak at about 288 eV is also caused by a core exiton[59-63].

## FULLERENE

Since their discovery, the spherical and hollow $C_{60}$ fullerenes have been fascinating subjects in solid state science due to their unique and useful physical properties such as a high electronic conductivity and superconductivity with a very high transition temperature. Therefore, a great deal of research has been performed to find out the electronic structures of the fullerene family ($C_{60}$ and its derivatives such as $C_{70}$) using photoemission spectroscopy (PES), x-ray absorption (XAS), EELS, and so forth. In this section, again EELS analyses will be the main part but to complete the understanding of the electronic response of fullerenes empirical results gained by other tools will be also mentioned briefly. Fullerenes for EELS study have been prepared and measured in the gas phase, in the solid phase on substrates, or in the form of an isolated solid film. In the solid state fullerenes form the face-centered cubic crystal structure as a result of the van der Waals cohesive interaction with a lattice parameter of 14.2 Å[64,65]. They show a phase transition to the simple cubic crystal structure at about 250 K. According to measurements by PES, inverse PES, and XAS with the gas and solid phases of $C_{60}$ fullerenes[66-70], those two physical phases of $C_{60}$ fullerenes have very similar electronic structures to each other though the solid phase shows some broadening of all energy levels compared to the gas phase. This demonstrates that in the solid phase $C_{60}$ molecules interact with each other very weakly. Also, from the number of excitations of $C_{60}$ molecules in the gas phase, the high degree of orbital degeneracy in $C_{60}$ fullerenes can be easily noticed. Namely, in spite of a large number of π electrons of $C_{60}$ (60 per molecule) the photonic and x-ray spectra present only a limited number of peaks.

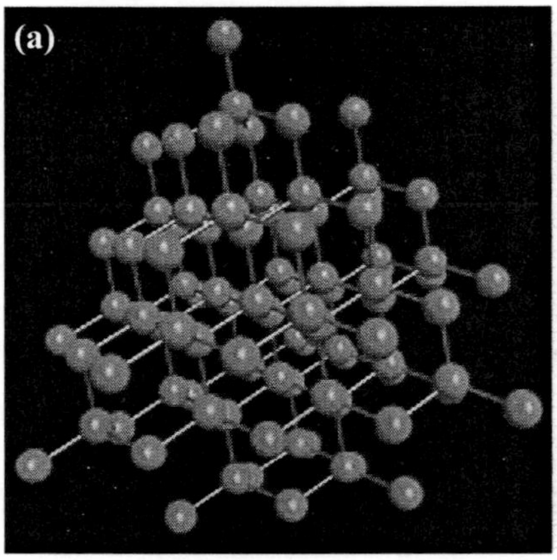


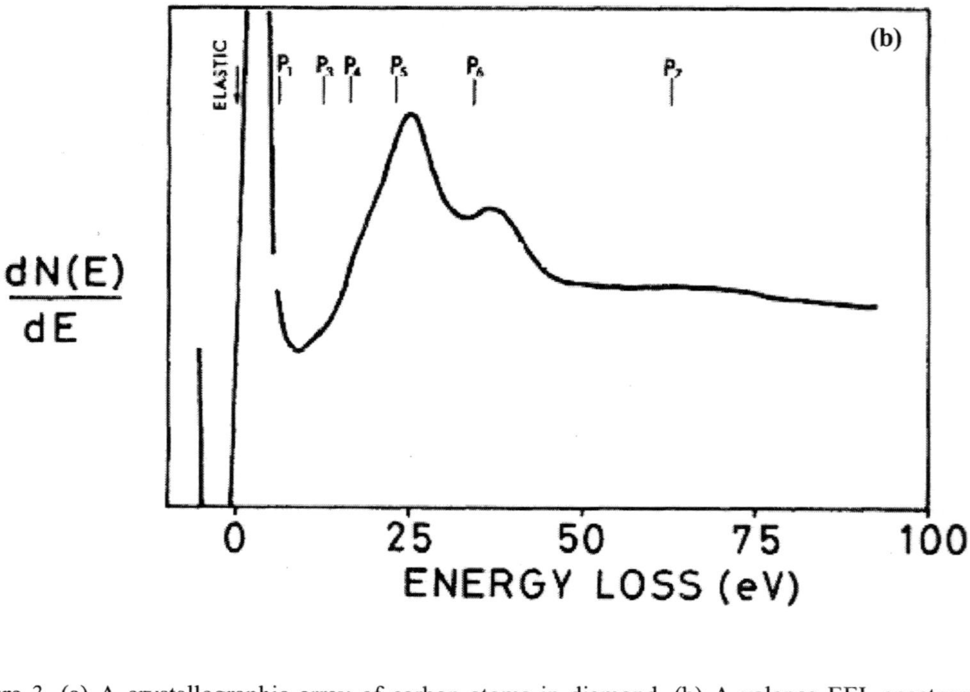


Figure 3. (a) A crystallographic array of carbon atoms in diamond. (b) A valence EEL spectrum of diamond (from reference 51).

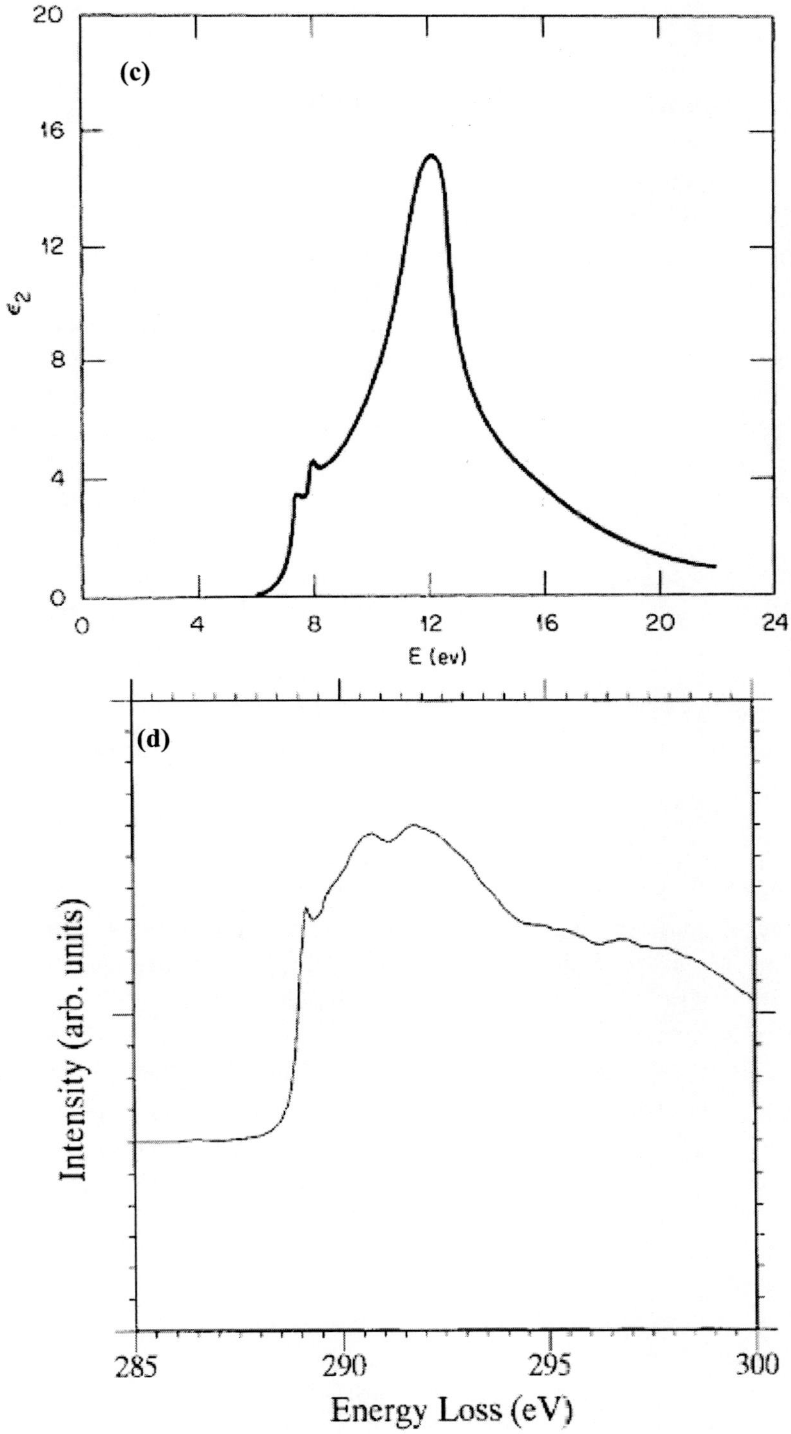


Figure 3. (c) An imaginary part of the dielectric function of diamond obtained by optical reflectance measurements (from reference 52). (d) A carbon K edge spectrum of diamond acquired by a high energy resolution STEM-attached EELS (from reference 58).

Figure 4 (a) – (c)[71,72] show valence EEL spectra taken from $C_{60}$ fullerite (containing less than 10% of $C_{70}$) on a silicon(100) substrate. The spectra were obtained using high resolution EELS in the reflection mode and the energy resolutions for the spectrum in Fig. 4 (a) through (c) are 0.015, 0.06, and 0.06 eV, respectively. The inset in Fig. 4 (a) displays a $C_{60}$ fullerene molecule in which a carbon pentagon is surrounded by five carbon hexagons and a carbon hexagon neighbors three pentagons and three hexagons. The EEL spectrum in Fig. 4 (a) obtained with the incident energy of 3.7 eV presents the excitations of the fullerite in the energy range corresponding to the infrared region. There are a few major peaks at 66, 94, 156, and 194 meV and additional minor peaks presented in the spectrum. According to the authors, the peak at 66 meV may be induced by strong dipole-active modes reinforced by the silicon substrate with a high dielectric constant in the infrared region. Furthermore, the strong Raman-active and radial-breathing $A_g$ mode at 496 $cm^{-1}$ is considered to cause the peak at 66 meV to be asymmetric. However, for the rest major peaks clear interpretations were not suggested. Instead, the authors related the excitations in the low energy loss spectrum of $C_{60}$ fullerite to the high densities of vibrational states around 58, 90 – 100, 155 – 175, and 180 – 200 meV of a graphite monolayer on transition metal carbide[73]. Therefore, the major peaks of $C_{60}$ fullerite obtained by high resolution EELS coincide with the four high density zone of a graphite monolayer, indicating the similarity between the electronic structures of the two carbon allotropes. The minor peaks in high energy region (around 225, 255, and 355 meV) in Fig. 4 (a) are assigned to the complex systems of the four strong peaks below 200 meV, that is, (66 + 156), (66 + 194 and 94 + 156), and (156 + 194 and 2 x 194) with the stretching motion of the C-H bond formed from contamination also contributing to the peak at 355 meV.

Figure 4 (b) shows a low loss spectrum up to 4 eV taken with a 10 eV primary energy exhibiting major peaks at 1.55, 2.2, and 3.7 eV and a few minor peaks. The peak at 1.55 eV is close to the $C_{60}$ fullerite band gap (1.5 eV)[74] and somewhat different from 1.7 eV for the highest occupied molecular orbital – lowest unoccupied molecular orbital level separation[75,76]. This peak shifts to higher energies as the primary energy increases, such that with a 30 eV primary energy the peak is seen at about 1.8 eV. The energy differences between the peak position and the band gap (and molecular orbital separation) along with its shift to higher energies stem from the fact that the peak at 1.55 indicates the energy necessary for generating the on-ball (or Frenkel) excitons[77-79]. On the other hand, the peak at 2.2 eV is assigned to an optically forbidden $h_u - t_{1u}$ transition across the fundamental gap[80]. The optical excitations of a material correspond to its response to photons with a momentum transfer close to zero (this is one reason that EELS, capable of tuning the momentum transfer delicately, is necessary for a complete understanding of the electronic structures of a material). By photons or electrons incident at small momentum transfers dipole-allowed transitions can be strongly activated whereas as the momentum transfer increases monopole- and quadrupole-allowed transitions also become strong in the spectra. Therefore, the peak at 2.2 eV is also considered to be a partially allowed transition[81] because it is seen as a shoulder when measured with a small momentum transfer. Its intensity becomes higher with increasing momentum transfers. The peaks at 3.7 and 4.8 (Fig. 4 (c)) eV are attributed to allowed ($h_u - t1_g$; $h_g$, $g_g - t1_u$) and $h_u - h_g$ transitions, respectively[80]. Basically, the peaks at 2.2, 3.7, and 4.8 eV are induced by molecular $\pi$ - $\pi$ interband transitions of $C_{60}$ fullerenes and slight differences about the peak positions are considered to be the result of energy resolution of each analyzing tool.

A very strong peak is seen at 6.3 eV in Fig. 4 (c) which is assigned to the collective π plasmon peak. One unique feature of $C_{60}$ fullerenes is that this π plasmon shows negligible dispersion with the momentum transfer[81,82]. This behavior is quite different from those of graphite[37] and single wall carbon nanotubes[83] with which the π plasmon disperses noticeably as the momentum transfer increases. While the peak positions of localized interband transitions are irrelevant to the momentum transfer, those of delocalized plasmons are expected to disperse with the momentum transfer. The minor peaks at around 10, 13, and 15 eV are primarily ascribed to the σ - σ* interband transitions with plausible secondary contributions from the π - σ* and σ - π* interband transitions[84]. The broad peak at 28 eV is assigned to the collective excitations of the all the 240 (π + σ) electrons of a $C_{60}$ fullerene and the σ plasmon[71,72,84]. The fact that the peak at 28 eV represents the molecular property of a $C_{60}$ fullerene rather than the solid state property of the $C_{60}$ fullerite also supports that the fullerite is formed by the weak van der Waals force. For the fullerite on a substrate, PES and EELS analyses showed that the electronic structure of the fullerite is not modified except the first layer of molecules in intimate contact with the substrate[85].

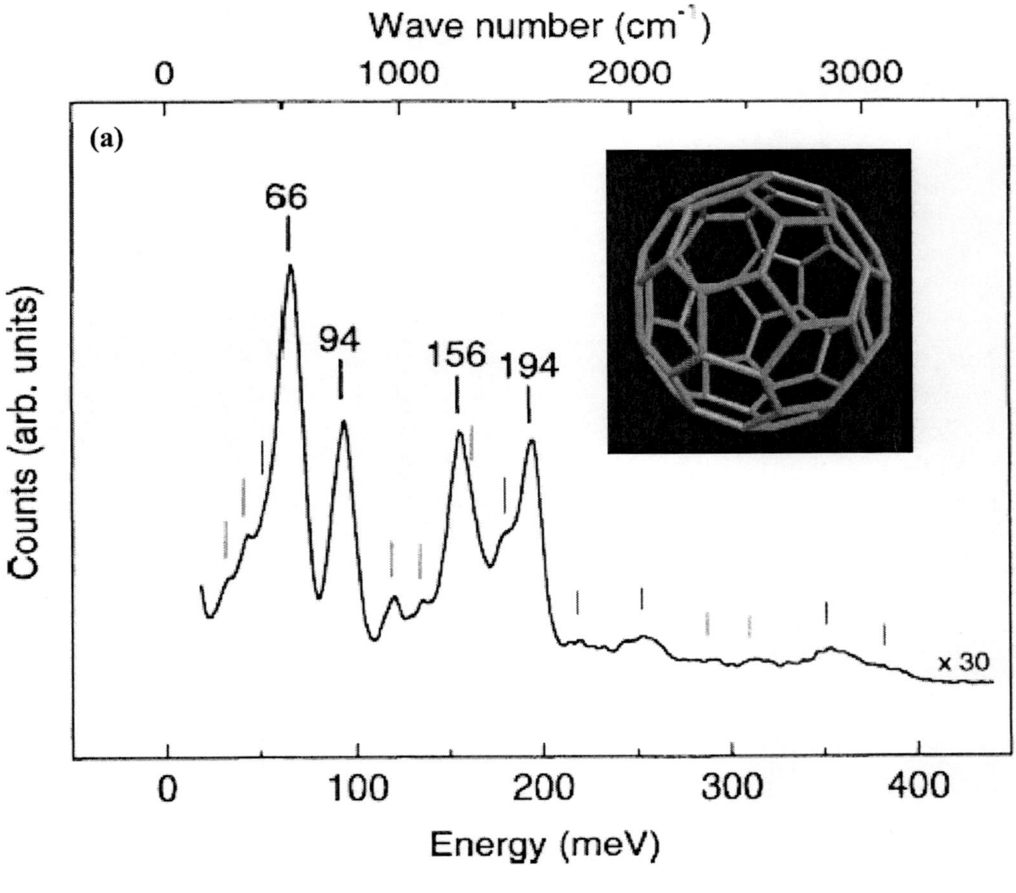


Figure 4 continued on next page

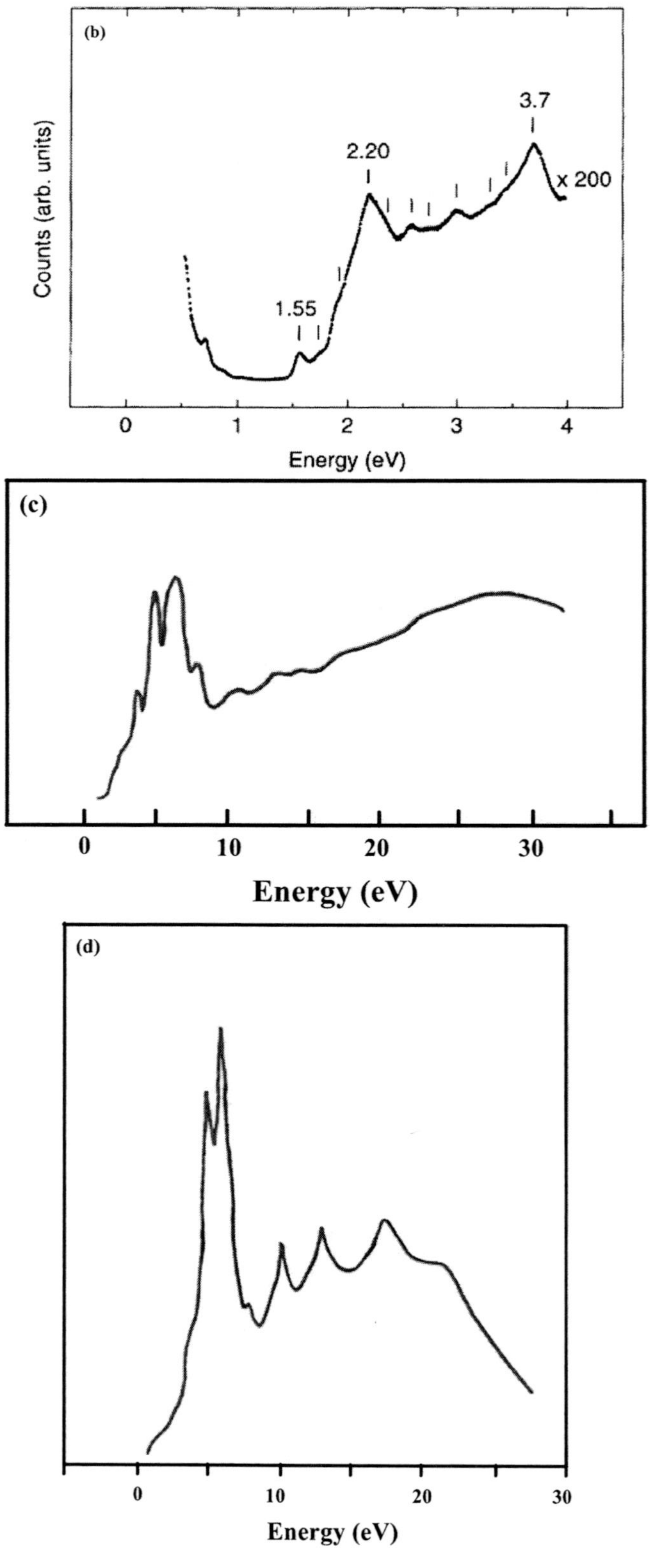


Figure 4 continued on next page

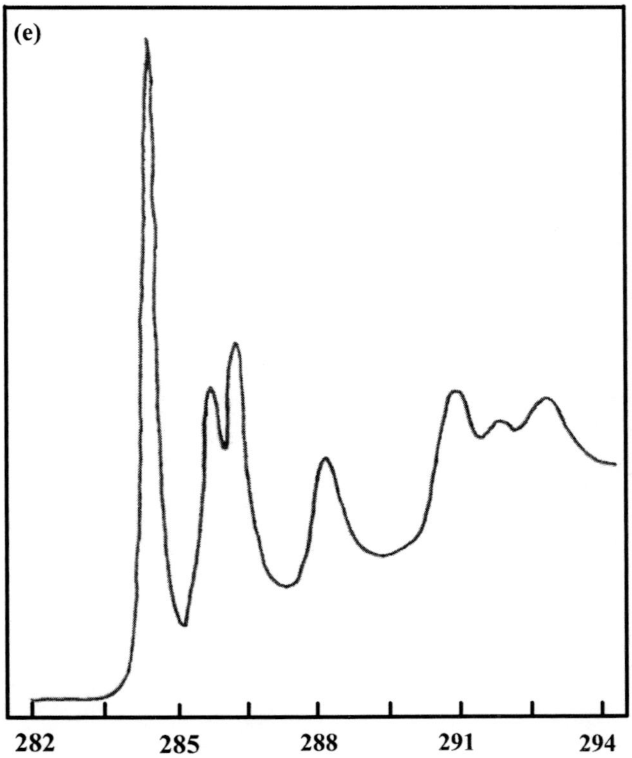


Figure 4. High resolution EEL spectra of a $C_{60}$ fullerite on a silicon (100) substrate corresponding to (a) the infrared region, (b) the near-infrared and visible region, and (c) the visible and ultraviolet region (from references 71 and 72). The inset in (a) shows the cage structure of a $C_{60}$ fullerene. (d) Valence EEL spectrum of $C_{60}$ fullerenes in the gas phase (from reference 86). (e) Core EEL spectrum of an isolated $C_{60}$ fullerite film (from reference 87).

$C_{60}$ fullerenes in the gas phase show a similar behavior in the valence EEL spectrum (Fig. 4 (d)) to the behavior of the solid fullerenes below 20 eV, indicating again that the $C_{60}$ molecules interact with each other weakly in the fullerite, that is, the van der Waals solid[86]. However, a remarkable difference is that the peak at 28 eV is not seen in the gas phase. The authors explain that this difference in the spectra may originate not only from the difference in the physical state but from the experimental conditions[86]. Therefore, the valence EEL spectrum in Fig. 4 (c) obtained with a relatively large momentum transfer also shows quadrupole or higher multipole transitions, which may lead to the appearance of the peak at 28 eV. There is also a large difference between the imaginary parts of dielectric functions for the gas and solid phases. Li *at al*[81] demonstrated that local field effects in the solid phase of the $C_{60}$ fullerite, like the case with graphite and diamond (see above), can make the difference in the dielectric functions.

Figure 4 (e) shows the core excitations of an isolated $C_{60}$ fullerite film[87]. One of the most outstanding features is that between the transitions from the carbon 1s orbitals to the $\pi^*$ or $\sigma^*$ states (~ 284 and 290 eV, respectively) a few additional transitions from carbon 1s to the other $\pi^*$ bands are shown at about 286, 286.5, and 288.5 eV. These peaks below 290 eV correspond to the unoccupied π* bands derived from $t_{1u}$, ($t_{2u}$ + $h_g$), and ($g_u$ + $g_g$ + $t_{2g}$) levels[74,80,88]. Another feature is that compared to the large width of the 1s to π* transitions of graphite, that of the $C_{60}$ fullerite is very small (~ 0.4 eV). Since the width of the carbon 1s to π* transitions show roughly the density of states of empty π*states, the small width for the transitions indicates the narrow bandwidth of the conduction π* band.

## Carbon Nanotubes

CNTs are the most recently-introduced allotrope of carbon. Their unique mechanical, electrical, and optical properties already found their uses in polymer-CNT composites, field emission display, and hydrogen storage for fuel cell[89]. They have a tubular shape rolled up seamlessly from graphitic layers. When CNTs consist of one graphitic layer, they are called single wall carbon nanotubes (SWNTs). MWNTs are composed of two or more graphitic layers. These CNTs have been intensively studied also by EELS to find out their electronic properties relatively in a short time. The band gap of SWNTs is known to be dependent on their chirality and diameter in contrast to that of boron nitride nanotubes that is independent of both variants. The response of SWNTs and MWNTs to the electron beam is different both in the low energy region and in the core excitation region. The upper right inset of Fig. 5 (a)[90] shows a SWNT where each carbon hexagon neighbors six other hexagons. SWNTs with this type of hexagon arrangement are called armchair SWNTs. The pentagon at the tip indicates a defect. Figure 5 (a)[83] shows the response of a purified SWNT film taken in a transmission mode with various momentum transfers by a dedicated EELS. These energy loss functions are obtained after removing the elastic peak and the contributions from the plural scattering as mentioned above. With a low momentum transfer, peaks at 0.85, 1.45, 2.0, 2.55, 3.7, 5.2 and 6.4 eV are seen. The first four peaks do not change their positions with the momentum transfer, indicating that these peaks stem from localized interband transitions of the π electron system in the SWNTs. The peak at 5.2 eV shows a large dependence on the momentum transfer, which is assigned to the π plasmon. However, the peaks at 3.7 and 6.4 eV become obscure with the momentum transfer though they also originate from the π electrons. In the upper left inset of Fig. 5 (a) a wide range of an energy loss function is shown, where the π + σ plasmon is found at 21.2 eV. This peak also disperses as the momentum transfer increases. The authors[83] relate the nature of the transitions to their polarization, so that the non-dispersive and localized interband transitions in the energy loss functions come from the excitations of energy states with the polarization perpendicular to the axis of CNTs while the dispersive and delocalized π plasmon stems from those of states polarized along the axis of CNTs. They also compared the dispersiveness of the plasmons for SWNTs and graphite, showing the close similarity between them and thus indicating the graphitic nature of SWNTs. In contrast to the case of $C_{60}$ fullerenes, the overall shapes of the energy loss functions of a SWNT film and the valence EEL spectra of an isolated SWNT molecule are quite different (Fig. 5 (b)[91] – compare with the inset in Fig. 5 (a)). In the valence EEL spectra of an isolated SWNT molecule taken by a STEM (VG HB501)-attached EELS, the π + σ plasmon peak at ~ 21 eV is not seen while a weak peak shows up at ~ 19 eV with a penetrating electron beam. Also, the excitation at ~ 15 eV stands out as a major peak and the π plasmon peak at ~ 5 eV is strongest. By comparing various inner and outer diameters of MWNTs, the authors assigned the 15 eV and the 19 eV peaks to tangential and radial modes, respectively, of surface plasmons. Reed and Sarikaya measured both SWNT bundles and molecules also with a STEM (VG HB501)-attached EELS[92]. However, the results are a bit different from the ones above partially because of a different deconvolution method and a different peak identification method. The deconvolution method[92] is that after subtracting the zero loss peak with a separately measured elastic peak, the rest part of the valence EEL spectrum was fitted by a few Lorentzian curves. They found the $\pi + \sigma$ plasmon at 23 – 27 eV and the $\pi$ plasmon at 4.5 – 5.5 eV in both a SWNT bundle and a SWNT molecule. Surface plasmons are observed in a rather wide range of 13 – 15 and 17 – 19 eV also in both the bundle and the molecule. The relative intensities of the two surface plasmons are different in the references 91 and 92. Namely, the lower-energy surface plasmon always shows larger intensity in the reference 91 but it displays smaller intensity with a penetrating electron beam in the reference 92. The valence EEL spectra in the reference 92 also show $\pi$ interband transitions at 2 – 3 eV in both the bundle and the molecule, which may not be possibly deconvoluted by the previously employed method. This will be discussed further when MWNTs are dealt with.

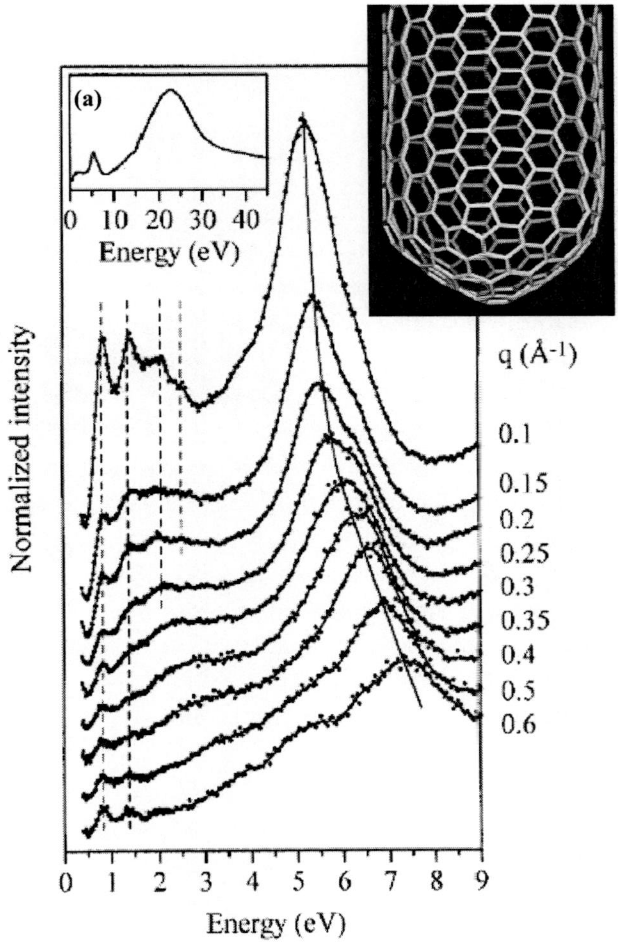


Figure 5 continued on next page

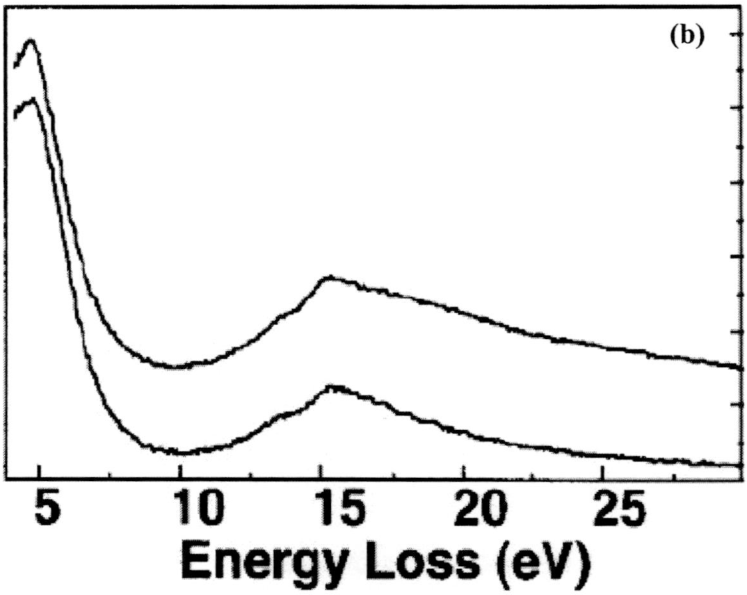


Figure 5. (a) Energy loss functions of a SWNT film taken by a dedicated EELS with various momentum transfers (from reference 83). The upper left inset shows an energy loss function of a wide energy range acquired with a low momentum transfer (0.15 $Å^{-1}$). The upper right inset displays an atomic array of a SWNT (from reference 90). (b) Low EEL spectra of an isolated SWNT obtained by an STEM-attached EELS with spatially-different beam positions (from reference 91). The upper spectrum is for a penetrating electron beam and the lower one is for a glancing electron beam.

In Fig. 6 (a) a purified MWNT with a diameter of about 50 nm is shown. The fringes indicate the walls of the nanotube. Also, the inset shows variations in the transmitted beam intensity along the smaller white line, so that the gap (about 0.35 nm) between the valleys stands for the interwall distance of the MWNT. The low loss spectra (Fig. 6 (b) – (d)) were obtained from different spatial positions (positions (b) – (d) in Fig. 6 (a)) after removing the zero loss peak and plural scattering through the Fourier-Log deconvolution. Figure 6 (b) shows two noticeable excitations of the delocalized and collective $\pi$ and $\pi + \sigma$ plasmons at 6.5 and 27 eV when measured with the electron beam penetrating the center of the MWNT (position (b) in Fig. 6 (a)). These peak positions are almost the same with those of graphite. This is because MWNTs of large diameters are very close to graphite in electronic properties[93,94]. Also, there is another peak at around 1.8 eV. This peak does not appear when a low loss spectrum taken at vacuum with no object under the same experimental conditions is deconvoluted through the same method. However, it is also shown from a low loss spectrum of a non-carbon material (ZnO) after deconvolution. Zinc and oxygen are known to show no valence transition below 3 eV. Therefore, this peak is at least partially an artifact of the deconvolution. On the other hand, there have been many theoretical and experimental reports about the localized and individual interband transitions of graphitic materials below 4 eV. For single wall CNT (SWNT) bundles these transitions appear at around 0.85, 1.45, and 2.0 eV involved with interband transitions between the van Hove singularities in the electronic density of states of metallic and semiconducting SWNTs when measured with a high energy resolution EELS[83], an optical spectrometer[95], and Rayleigh scattering[96]. Also, by optical measurements MWNTs showed the interband transitions by the van Hove singularities at similar excitation energies[97]. For highly oriented pylorytic graphite a transition is seen at 1.2 eV as a shoulder in a high energy resolution EELS[19]. From another report on SWNTs the π - π* interband transition dominates at 2 – 3 eV in valence EEL spectra[92]. Also, from *ab initio* calculations, it was reported that the π - π* interband transitions of graphite are responsible for the absorption peaks (or the peaks in the imaginary part of the dielectric function) below 4 eV as mentioned above. The peak at about 1.8 eV increases greatly as the probe moves to the edge (Fig. 6 (b) – (d)). This spatial dependence of the excitation is just as expected from the π - π* interband transitions[92]. However, a great caution should be used when data points below 3 eV in EEL spectra obtained by STEM-attached EELS are interpreted. Electrons from the electron gun of STEM go through different paths when they penetrate the different spatial points of the CNT in Fig. 6 (a). Therefore, electrons passing the point (b) experience the longest path whereas as the probe moves to the edge the penetration path gets smaller. The difference in the electron paths reflects that in the intensity of the zero loss peak. The ratio of the zero loss peaks in Fig. 2 (b) through (d) is approximately similar to that of the peak at 1.8 eV of the EEL spectra. Therefore, the increase in the intensity of the peak at 1.8 eV by the movement of the electron beam to the edge of the CNT can be attributed to the increase in the intensity of the elastic scattering due to the shorter path. As mentioned above, this peak was not seen with a zero loss peak taken at a vacuum but observable with a non-carbon material. Therefore, even though the peak at 1.8 eV shows a very similar trend to the π - π* interband transitions determined in the reference 92, it is now defined only as an artifact of the Fourier-Log deconvolution. In the reference 92 they used their own devised deconvolution method to sort out the peak at 2 – 3 eV. The authors emphasize that the peak at 2 – 3 eV was not observed when the reference background spectrum was deconvoluted and thus it should be assigned to the π - π* interband transitions. However, since the peak at 1.8 eV was also not shown from the reference spectrum processed by the deconvolution method in current study, there is still a possibility that in their reasoning the effect of the path difference might be ignored. In the data they used, the π - π* interband transitions increase with the electron beam shifting to the edge of a SWNT bundle while in the case of an independent SWNT their intensity is almost the same with the beam shift. This would be exactly the same trend with the elastic zero loss peaks of a SWNT bundle and a very small diameter SWNT. There is a slight difference (larger than 0.2 eV) in the peak position between current and their results[92], which may be induced from the differences in the deconvolution methods.

Figure 6 (b) through (d) show the spatial dependence of the excitation probability of the modes obtained in current study. By placing the electron probe from the center to the edge of the nanotubes, the excitations of the π and π + σ plasmons decay with their peak positions shifting to lower energies and new excitations show up at about 13 and 17 eV (the arrows). This result exhibits a very similar trend with previous report[98]. The authors in the reference interpret the shift of the π plasmon with the spatial points as a competition of the π bulk plasmon (6 – 7 eV) and the π surface plasmon (6 eV). Therefore, as the electron beam moves to the edge, the π surface plasmon becomes more important, shifting the peak position to lower energies. In the case of the π + σ plasmon, they used the spatial preference of two bulk modes to explain its peak shift. Namely, the π + σ plasmon is, in fact, composed of two bulk modes; one is at 23 eV, which is easily excited when the graphene layers are parallel to the electron beam and the other is at 27 eV, which is, in contrast to the 23-eV mode, dominant when the layers are perpendicular to the electron beam. As a result, the relative contribution of the two bulk plasmons varies when the electron beam moves, explaining the shift of the $\pi + \sigma$ plasmon. They also assigned the peak at 13 eV to the $\pi$ - $\sigma^*$ and $\sigma$ - $\pi^*$ surface plasmons, and the peak at 17 eV to the $\sigma$ surface plasmon subject to shift also with the beam position.

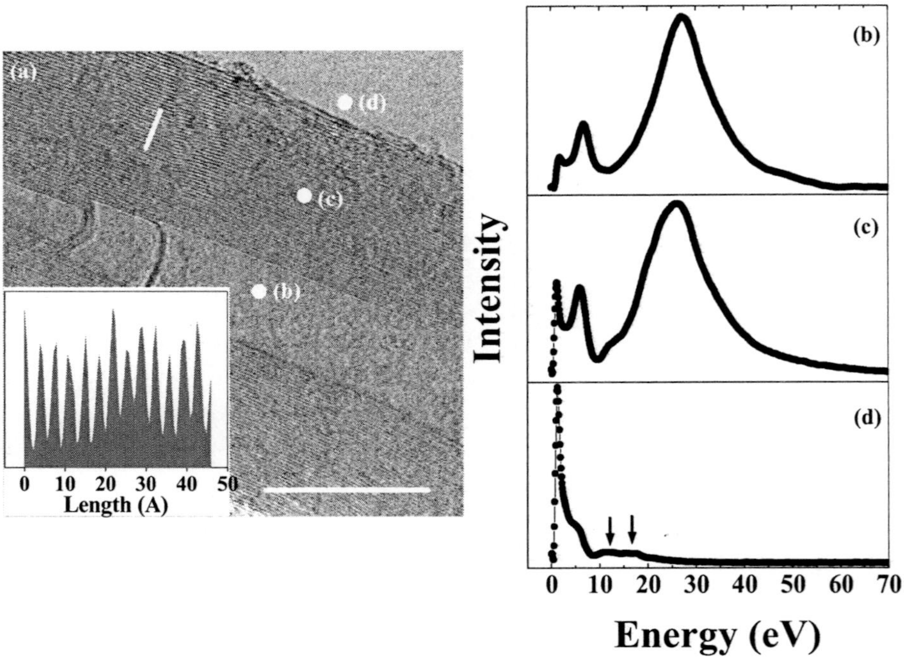


Figure 6. (a) A TEM image of a MWNT. The inset shows the line scan along the smaller white line. The average distance between the valleys is about 0.35 nm very close to the interlayer distance of graphite. The scale bar (larger white line) is 20 nm. (b) – (d) Valence EEL spectra obtained from different positions ((b) – (d)) of the MWNT in (a). The $\pi + \sigma$ plasmon at ~ 27 eV and the $\pi$ plasmon at ~ 6.5 eV are clearly seen in Fig. (b) and (c) while in Fig. (c) the $\pi + \sigma$ plasmon disappears and the $\pi$ plasmon is partially embedded in the false peak at ~ 1.8 eV. The arrows in (d) indicate peaks at about 13 and 17 eV.

It was reported that the peak positions for the $\pi$ and $\pi + \sigma$ plasmons shift to lower energies as the diameter of MWNTs decreases[99,100]. For MWNTs of a large diameter, the peak positions of the $\pi$ and $\pi + \sigma$ plasmons are as seen above. Ajayan *et al*[99] explains that the peak shift of the $\pi + \sigma$ plasmon is the consequence of a curvature effect in MWNTs while Stockli *et al*[100] ascribe it to the relatively large contribution of surface plasmons in MWNTs of small diameters. However, according to the reference[98], it might be also deduced that as the diameter of a MWNT decreases the $\pi$ surface plasmon mode and the 23-eV bulk plasmon mode become more important since the electrons penetrating the center area of a small-diameter nanotube have a higher chance of seeing the edge simultaneously. Also, according to the references[20,101,102], the plasmon energy is proportional to the square root of the density of valence electrons, so that the shift of the π and π + σ plasmons to lower energies for smaller-diameter MWNTs may be interpreted as the decrease in the density of valence electrons. In this context, the result of Dravid *et al*[103] may not be clearly understood since in their displayed EEL spectra the π plasmon is seen at a lower energy for a MWNT than for graphite while the π + σ plasmon is observed almost at the same energy for both materials.

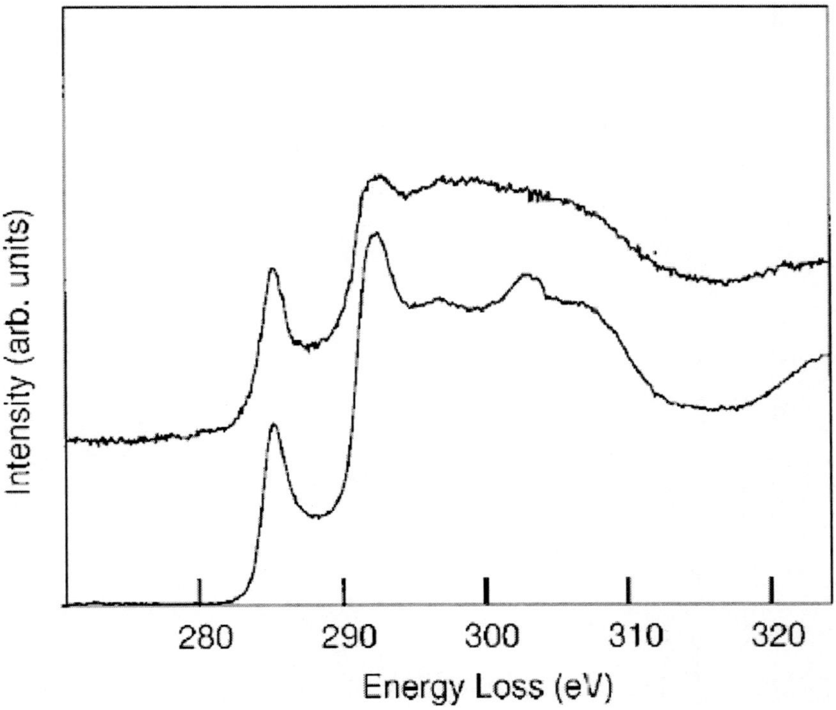


Figure 7. Carbon K edge spectra of a SWNT with a diameter of 1 nm (upper) and a MWNT with a diameter of 30 nm (lower) taken by an STEM-attached EELS (from reference 104).

Core loss spectra of a SWNT and a MWNT are seen in Fig. 7[104]. The carbon K edges of both nanotubes (285.5 and 291 eV, respectively) are found almost at the same energies with those of graphite. The authors of the reference also show that the π* and σ* core transitions shift to lower energies when the electron beam moves from the center to the edge of a MWNT. The spectrum structures above 291 eV of the MWNT are different from those of the SWNT but close to those of graphite. The SWNT does not have any sharp peak but a broad resonance. On the other hand, the spectrum structures above 291 eV of the MWNT shows three more peaks at the energies similar to those of graphite. Ajayan *et al*[99] reported a broadening and shift of the π* peak, and shift of the σ* peak for small-diameter MWNTs. However, in the Fig.7[104] even the 1-nm-diameter SWNT does not show any broadening or shift to higher energies of the peaks compared to the 30-nm-diameter MWNT. The authors[104] believe that amorphous carbon on CNTs may affect the core EEL spectra of the reference 99, which can broaden the π* peak. Dravid *et al*[103] also compared C K edges of a MWNT and graphite, and showed that when the electron beam is incident perpendicular to the axis of the MWNT and parallel to the c-axis of graphite, the π*/σ* ratio of graphite is larger than 1 whereas that of the MWNT is smaller than 1. They attributed the difference to a greater σ* contribution of a MWNT by the curvature of the nanotube. To explain more in detail, this π*/σ* ratio value of graphite corresponds to the experimental conditions of small convergence and collection angles (that is, most incident electrons into and out of graphene layers are parallel to the c-axis of graphite) as seen above in the graphite section. Under these conditions, the graphene layers at the edge of the nanotube stand parallel to the electron beam, satisfying the dielectric selection rule for the 1s to σ* transition and thus strengthening the σ* intensity.

## REFERENCES


[1] *http://en.wikipedia.org/wiki/Allotropes_of_carbon*
[2] R. Setton, P. Bernier, S. Lefrant, *Carbon Molecules and Materials* (CRC, New York, 2002)
[3] A. Hirsch, *Topics Curr. Chem.* 199, 1 (1999)
[4] C. Kittel, *Introduction to Solid State Physics* 6th Ed. (Wiley, New York, 1991)
[5] S. Saito and A. Oshiyama, *Phys. Rev. Lett.* 66, 2637 (1991)
[6] M. S. Dresselhaus, G. Dresselhaus, P. C. Eklund, *Science of Fullerenes and Carbon Nanotubes: Their Properties and Applications* (Academic Press, Orlando, 1996)
[7] J. W. G. Wilder, L. C. Venema, A. G. Rinzler, R. E. Smalley and C. Dekker, *Nature* 391, 59 (1998)
[8] L. Henrard, A. A. Lucas, and Ph. Lambin, *Astrophys. J.* 406 92 (1993)
[9] L. Henrard, Ph. Lambin, and A. A. Lucas, *Astrophys. J.* 487, 719 (1997)
[10] L. Henrard, and Ph. Lambin, *J. Phys. B* 29, 5127 (1996)
[11] G. Binnig and H. Rohrer, *Rev. Mod. Phys.* 59, 615 (1987).
[12] J. Golovchenko, *Science* 232, 48 (1986).
[13] B. K. Teo and D. C. Joy, *EXAFS spectroscopy: Techniques and Applications* (Plenum, New York, 1981)
[14] P. R. Griffiths and J. A. de Haseth, *Fourier Transform Infrared Spectrometry* (Wiley, New York, 1986)
[15] D. A. Long, *Raman Spectroscopy*, (McGraw-Hill, New York, 1977)
[16] R. E. Collins, *Rev. Sci. Instr.* 61, 2029 (1990)
[17] A. V. Crewe, M. Isaacson, and D. E. Johnson, *Rev. Sci. Instr.* 42, 411 (1971)
[18] R. F. Egerton, *Electron Energy-Loss Spectroscopy in the Electron Microscopy*, 2nd Ed. (Plenum, New York, 1996)
[19] J. J. Ritsko and E. J. Mele, *Phy. Rev. B* 21, 730 (1980)
[20] D. Pines, *Elementary Excitations in Solids* (Benjamin, New York, 1964)
[21] H. Kim, M. Kaufman, W. Sigmund, D. Jacques, and R. Andrews, *J. Mater. Res.* 18, 1104 (2003)
[22] R. Andrew, D. Jacques, D. Qian and E. C. Dickey, *Carbon*, **39**, 1681, (2001)
[23] D. B. Williams and C. B. Carter, *Transmission Electron Microscopy* (Plenum, New York, 1996)
[24] A. J. Dekker, *Electrical Engineering Materials* (Prentice-Hall, New Delhi, 1973)
[25] D. G. McCullough, D. R. McKenzie and C. M. Goringe, *Phy. Rev. B* 621, 2349 (2000)

[26] S. D. Berger and D. R. McKenzie, *Philos. Mag. Lett.* 57, 285 (1988)

[27] A. J. Papworth, C. J. Kiely, A. P. Burden, S. R. P. Silva, and G. A. J. Amaratunga, *Phy. Rev. B* 62, 12628 (2000)

[28] Hamon, J. Verbeeck, D. Schryvers, J. Benedikt and R. M. C. M. v. d. Sanden, *J. Mater. Chem.* 14, 2030 (2004)

[29] P. Hitchcock, D. C. Newbury, I. Ishii, J. A. Horsley, R. D. Redwing, A. L. Johnson, and F. Sette, *J. Chem. Phys.* 85, 4849 (1986)

[30] J. Yuan, L. M. Brown, *Micron*, 31, 515 (2000)

[31] F. Sette, J. Stohr, and A. P. Hitchcock, *J. Chem. Phys.* 81, 4906 (1984)

[32] R. McLaren, S. A. C. Clark, I. Ishii, and A. P. Hitchcock, *Phy. Rev. A* 36, 1683 (1987)

[33] Rafferty and L. M. Brown, *Phy. Rev. B* 58, 10326 (1998)

[34] Alexandrou, A. J. Papworth, B. Rafferty, G. A. J. Amaratunga, C. J. Kiely, L. M. Brown, *Ultramicroscopy* 90, 39 (2001)

[35] E. A. Taft, and H. R. Philipp, *Phys. Rev.* 138, A197 (1965)

[36] K. Zeppenfeld, *Z. Phys.* 211, 391 (1968)

[37] K. Zeppenfeld, *Z. Phys.* 243, 229 (1971)

[38] L. Greenaway, G. Harbeke, F. Bassani, and E. Tosatti, *Phys. Rev.* 178, 1340 (1969)

[39] Y. Ma, P. Skytt, N. Wassdahl, P.Glans, D. C. Mancini, J. Guo, and J. Nordgren, *Phys. Rev. Lett.* 71, 3725 (1993)

[40] P. A. Bruhwiler, A. J. Maxwell, C. Puglia, A. Nilsson, S. Andersson, and N. Martensson, *Phys. Rev. Lett.* 74, 614 (1995)

[41] R. Ahuja, S. Auluck, J. M. Wills, M. Alouani, B. Johansson, and O. Eriksson, *Phys. Rev. B* 55, 49999 (1997)

[42] M. Vos, A. S. Kheifets, E. Weigold, and F. Aryasetiawan, *Phys. Rev. B* 63, 033108 (2001)

[43] A. G. Marinopoulos, L. Reining, V. Olevano, A. Rubio, T. Pichler, X. Liu, M. Knupfer, and J. Fink, *Phys. Rev. Lett.* 89, 076402 (2002).

[44] J. F. Shackelford. *Introduction to Materials Science for Engineers* (2$^{nd}$ Ed.). (Macmillan, New York, 1990)

[45] Y. X. Zhao and I. L. Spain, *Phys. Rev. B* 40, 993 (1989)

[46] N. K. Menon, and J. Yuan, *Ultramicroscopy* 74, 83 (1998)

[47] R. A. Rosenberg, P. J. Love, and V. Rehn, *Phys. Rev. B* 33, 4034 (1986)

[48] J. Yuan, and N. K. Menon, *J. Appl. Phys.* 81, 5087 (1997)

[49] H. Raether, *Springer Tracts in Modern Physics*, edited by G. Hohler (Springer, Berlin, 1965), Vol. 38, p. 85

[50] A. G. Marinopoulos, L. Reining, A. Rubio, and V. Olevano, *Phys. Rev. B* 69, 245419 (2004)

[51] H. Armon, and J. P. F. Sellschop, *Phys. Rev. B* 26, 3289 (1982)

[52] G. S. Painter, D. E. Ellis, and A. R. Lubinsky, *Phys. Rev. B* 4, 3610 (1971)

[53] R. A. Roberts, and W. C. Walker, *Phys. Rev* 161, 730 (1967)

[54] J. Himpsel, J. F. van der Veen, and D. E. Eastman, *Phys. Rev. B* 22, 1967 (1980)

[55] G. Lurie, and J. M. Wilson, *Surf. Sci.* 65, 476 (1977)

[56] H. G. Maguire, *Phys. Status Solidi B* 76, 715 (1976)

[57] S. Waidmann, M. Knupfer, B. Arnold, J. Fink, A. Fleszar, and W. Hanke, *Phys. Rev. B* 61, 10149 (2000)

[58] M. Terauchi, M. Tanaka, K. Tsuno, and M. Ishida, *J. Microsc.* 194, 203 (1999)

[59] J. F. Morar, F. J. Himpsel, G. Hollinger, G. Hughes, and J. L. Jordan, *Phys. Rev. Lett.* 54, 1960 (1985)
[60] K. A. Jackson, and M. R. Pederson, *Phys. Rev. Lett.* 67, 2521 (1991)
[61] J. Nithianandam, *Phys. Rev. Lett.* 69, 3108 (1992)
[62] P. E. Batson, *Phys. Rev. Lett.* 70, 1822 (1993)
[63] Y. Ma, P. Skytt, N. Wassdahl, P. Glans, D. C. Mancini, J. Guo, and J. Nordgren, *Phys. Rev. Lett.* 71, 3725 (1993)
[64] R. M. Fleming, T. Siegrist, P. M. Marsh, B. Hessen, A. R. Kortan, D. W. Murphy, R. C. Haddon, R. Tycko, G. Dabbagh, A. M. Mujsce, M. L. Kaplan, and S. M. Zahurac, *Mater. Res. Soc. Symp. Proc.* 206, 691 (1991)
[65] P. A. Heiney, J. E. Fischer, A. R. McGhie, W. J. Romanow, A. M. Denenstein, J. P. McCauley, A. B. Smith, and D. E. Cox, *Phys. Rev. Lett.* 66, 2911 (1991)
[66] L. Lichtenberger, M. E. Jatko, K. W. Nebesny, C. D. Ray, D. R. Huffman, and L. D. Lamb, *Mater. Res. Soc. Symp. Proc.* 206, 673 (1991)
[67] P. A. Bruhwiler, A. J. Maxwell, P. Rudolf, C. D. Gutleben, B. Wastberg, and N. Martensson, *Phys. Rev. Lett.* 71, 3721 (1993)
[68] S. Krummacher, M. Biermann, M. Neeb, A. Liebsch, and W. Eberharbt, *Phys. Rev. B* 48, 8424 (1993)
[69] M. R. C. Hunt, and P. Rudolf, S. Modesti, *Phys. Rev. B* 55, 7889 (1997)
[70] M. Pedio, M. L. Grilli, C. Ottaviani, M. Capozi, C. Quaresima, P. Perfetti, P. A. Thiry, R. Caudano, and P. Rudolf, *J. Electron Spectrosc. Related Phenom.* 76, 405 (1995)
[71] G. Gensterblum, J. J. Pireaux, P. A. Thiry, R. Caudano, J. P. Vigneron, A. A. Lucas, and W. Kratschmer, *Phys. Rev. Lett.* 67, 2171 (1991)
[72] A. Lucas, G. Gensterblum, J. J. Pireaux, P. A. Thiry, R. Caudano, J. P. Vigneron, Ph. Lambin, and W. Kratschmer, *Phys. Rev. B* 45, 13694 (1992)
[73] T. Aizawa, R. Souda, S. Otani, Y. Ishizawa, and C. Oshima, *Phys. Rev. Lett.* 64, 768 (1990)
[74] S. Saito and A. Oshiyama, *Phys. Rev. Lett.* 66, 2637 (1991)
[75] J. W. Mintmire, B. I. Dunlap, D. W. Brenner, R. C. Mowrey, and C. T. White, *Phys. Rev. B* 43, 14281 (1991)
[76] I. Dunlap, D. W. Brenner, J. W. Mintmire, R. C. Mowrey, and C. T. White, *J. Phys. Chem.* 95, 8737 (1991)
[77] R. W. Lof, M. A. van Veenendal, B. Koopmans, H. T. Jonkman, and G. A. Sawatzky, *Phys. Rev. Lett.* 68,3924 (1992)
[78] L. Shirley, L. X. Benedict, and S. G. Louie, *Phys. Rev. B* 54, 10970 (1996)
[79] R. Eder, A. M. Janner, and G. A. Sawatzky, *Phys. Rev. B* 53, 12786 (1996)
[80] Sohmen, J. Fink, and W. Kratschmer, *EuroPhys. Lett.* 17, 51 (1992)
[81] Li, S. Velasquez, and S. E. Schntterly, *Phys. Rev. B* 49, 2969 (1994)
[82] Romberg, E. Sohmen, M. Merkel, M. Knupfer, M. Alexander, M. S. Golden, P. Adelmann, T. Pietrus, J. Fink, R. Seemann, and R. L. Johnson, *Synth. Met.* 56, 3038 (1993)
[83] T. Pichler, M. Knupfer, M. S. Golden, J. Fink, A. Rinzler, and R. E. Smalley, *Phys. Rev. Lett.* 80, 4729 (1998)
[84] Sohmen and J. Fink, *Phys. Rev. B* 47, 14532 (1993)
[85] M. R. C. Hunt, P. Rudolf, S. Modesti, *Phys. Rev. B* 55, 7882 (1997)
[86] J. W. Keller and M. A. Coplan, Chem. *Phys. Lett.* 193, 89 (1992)

[87] M. S. Golden, M. Knupfer, J. Fink, J. F. Armbruster, T. R. Cummins, H. A. Romberg, M. Roth, M. Sing, M. Schmidt, and E. Sohmen, *J. Phys.: Condens. Matter* 7, 8219 (1995)
[88] J. L. Martins, N. Troullier, J. H. Weaver, Chem. *Phys. Lett*. 180, 457 (1991)
[89] S. Nalwa, *Encyclopedia of Nanoscience and Nanotechnology*, (American Scientific Publishers, New York, 2004)
[90] *http://smalley.rice.edu/smalley.cfm?doc_id=4866*
[91] O. Stephan, D. Taverna, M. Kociak, K. Suenaga, L. Henrard, and C. Colliex, *Phys. Rev. B* 66, 155422 (2002)
[92] B. W. Reed and M. Sarikaya, *Phys. Rev. B* 64, 195404 (2001)
[93] V. Likodimos, S. Glenis, N. Guskos, and C. L. Lin, *Phys. Rev. B* 68, 045417 (2003)
[94] L. A. Bursill, P. A. Stadelmann, J. L. Peng, and S. Prawer, *Phys. Rev. B* 49, 2882 (1994)
[95] O. Jost, A. A. Gorbunov, W. Pompe, T. Pichler, R. Friedlein, M. Knupfer, M. Reibold, H. D. Bauer, L. Dunsch, M. S. Golden, and J. Fink, *Appl. Phys. Lett*. 75, 2217 (1999)
[96] M. Y. Sfeir, F. Wang, L. Huang, C. Chuang, J. Hone, S. P. O'Brien, T. F. Heinz, and L. E. Brus, *Science* 26, 1540 (2004)
[97] M. E. Brennan, J. N. Coleman, A. Drury, B. Lahr, T. Kobayashi, and W. J. Blau, *Opt. Lett*. 28, 266 (2003)
[98] M. Kociak, L. Henrard, O. Stephan, K. Suenaga, and C. Colliex, *Phys. Rev. B* 61, 13936 (2000)
[99] P. M. Ajayan, S. Iijima, and T. Ichihashi, *Phys. Rev. B* 47, 6859 (1993)
[100] T. Stockli, J. Bonard, P. Stadelmann, and A. Chatelain, *Z. Phys*. D 40, 425 (1997)
[101] Williams and J. W. Edington, *J. Microsc* 108, 113 (1976)
[102] M. C. Cheynet, C. Barioz, S. Blais, and D. Duly, *Microsc. Microanal. Microstruct*. 5, 91 (1994)
[103] V. P. Dravid, X. Lin, Y. Wang, X. K. Wang, A. Yee, J. B. Ketterson, and R. P. H. Chang, *Science* 259, 1601 (1993)
[104] O. Stephan, P. M. AjayanC.Colliex, F. Cyrot-Lackmann, and E. Sandre, *Phys. Rev. B* 53, 13824 (1996)

In: Trends in Crystal Growth Research
Editor: George V. Karas, pp. 161-177

ISBN 1-59454-541-3


***Chapter 7***

# STUDY OF ICE NUCLEATION/ HYDRATE CRYSTALLIZATION OVER UREA

***A. Hazra[1], S. Paul[1], U. K. De[1], S. Kar[2] and K. Goswami[2*]***
[1]Atmospheric Science Research Group, Environmental Science Programme, Faculty of Science, Jadavpur University, Kolkata - 700 032.
[2]Dept. of Physics, Jadavpur University, Kolkata - 700 032.

## ABSTRACT

Aqueous solution of Urea is used as seeding agent for ice nucleation in a walk-in cold chamber. Ice nucleation is studied over a range of -19.0 °C to 0 °C. The variation of crystal concentration with temperature has a sharp peak at -11.9 °C for a specific Total Water Content (TWC). Incidentally, -11.9 °C coincides with the eutectic temperature of Urea and water mixture. When seeding is done by aqueous solution of Urea, there is a temperature drop and it has a variational pattern with seeding temperature, which is in conformity with the variation of crystal count. At fixed temperature ice crystallization depends on TWC and the number count attains a saturation value with rise in TWC. The aqueous solution of Urea has the ability to form hydrate crystal on the positive side of 0 °C up to +15.2°C. The maximum nucleation is found to occur at +11 °C. The hydrate nature of the crystal is confirmed after carrying out different supporting experiments. Hydrate crystals are grown in an incubator within the above temperature range. X-ray Diffraction (XRD) analysis establishes the presence of Urea as well as hydrated form in the grown crystal. Lastly, $^1$H-NMR study confirms the presence of Urea and water in the grown crystal.

**Key Words:** Nucleation, Eutectic temperature, Crystal growth, X-ray diffraction, Nuclear Magnetic Resonance

* Corresponding author (Prof. K. Goswami): Fax: +91-33-2413-8917 Tel No: +91-33-2414 6110; E-mail address: goswami_k2001@yahoo.com

# 1. INTRODUCTION

Ice nucleation and subsequent growth is an important aspect of cloud physics. Whether the water vapour or liquid water present in the atmosphere will convert into ice, is decided by a number of factors. It is well known that below the freezing point of water and under normal conditions of pressure, ice is more stable as compared to liquid water. On the other hand, if the vapour pressure exceeds the saturation value in an environment below 0°C, conversion to an ice phase is more stable than that in the vapour phase. In spite of being a more stable phase, ice cannot be spontaneously produced, as creation of ice embryos inside the parent phase needs additional energy. Only when ice has grown above a critical size, can it freely grow in a supercooled or supersaturated condition. Otherwise, water continues to remain in a supercooled state or as a vapour in its supersaturated state. However, the nucleation can be initiated by introduction of foreign particulate or surface and this gives rise to heterogeneous nucleation. The particulate may either be a solid particle or droplet. In fact, most of the nucleation in nature is heterogeneous in character. These foreign particulates, which help nucleate ice, are more commonly known as ice-forming nuclei (IN). Although, all foreign particulates cannot behave as ice nuclei (IN), some are good IN. The major factors determining the ability of ice nucleation are [1], lattice or structural match between the substrate and ice crystal, availability of hydrogen bond at the IN surface, size of the nucleant, the electric field present in the substrate that orients the water molecules and the presence of active sites on the surface of the substrate where water molecules are preferentially adsorbed. The physical parameters viz. temperature, vapour pressure and also the atmospheric pressure are crucial to nucleation. However, the exact relationship between ice crystal concentration and atmospheric condition is not clearly understood. Water insolubility of IN possibly enhances the stability. However, water soluble particles like NaCl, $(NH_4)_2SO_4$ act as reasonably good IN.

Historically the first IN was identified in late forties of the last century [2]. Some researchers carried out field experiments to collect ice crystals and could identify that atmospheric aerosols like clay mineral and dust particles form the majority of IN particles [3],[4],[5],[6]. Particulates generated during volcanic eruptions were also identified as source of IN [7][8]. Besides, various hygroscopic particles, particulates from combustion emissions and organic materials were also identified as possible IN. Many oxides which are mainly emitted during industrial activities, were also identified as IN in the atmosphere [9],[10][11][12][13].

Since the discovery of glaciogenic characteristics more than fifty years ago, both silver iodide and dry ice are still the most widely used cloud seeding materials in the world. Both materials enhance the ice crystal concentrations in clouds by either nucleating new crystals or freezing cloud droplets. Based on their ice-nucleating capabilities two seeding concepts have been proposed in the past, namely, the static and dynamic seeding concepts [14].

The natural sources can be separated into inorganic and organic. Until the late 1950s, it was thought that only inorganic materials could be responsible for initiating ice formation in the atmosphere, and also on vegetation and poikilotherm animals. Since late fifties of the last century, the importance of organic materials as ice nuclei (IN) gradually appeared and a good number of such substances were reported even in early sixty [15]. In fact, Phloroglucinol was identified as the first organic IN [16] and its ice nucleating behaviour has been studied in details in a recent article [17]. The Benzoin dust and alcoholic solution of Benzoin are also promising ice nucleating agent [18]. The Benzoin dust has the ability to form hydrate crystals above 0 °C [18]. During the following years, ice nucleation by organic materials became the subject of many investigations [19],[20],[21],[22] with interest in fundamental study as well as in practical field. The studies of ice nucleation by organic compounds have brought out the necessity for a reevaluation of the theory of the heterogeneous nucleation mechanism and became important in the search for an understanding of this process on the molecular level. From a practical point of view, these effective organic substances have appeared to cloud physicists as very promising cloud seeding agent due to their wide variety and effectiveness.

Among the organic and biogenic compounds naturally occurring in the atmosphere mention may be made of some plant litters, bacteria, lichen, fungi, mollusks, some freeze tolerant insects [23] and also some amino acids [24], which act as IN. Size of biogenic IN is quite wide, starting from a single protein on bacterial cell to a large nucleator of the size of several microns. It is claimed that [25], nearly 25% of all aerosols, which act as IN, are biogenic in character. Some amino acids such as L-leucine were found to possess ice nucleating ability with a threshold temperature as high as - 4.5°C [24],[26],[27]. However, until now, the exact specific origin of the atmospheric amino acids has not been known with certainty. Most likely, these are of biological origin. Recently, Gorzelska [28] identified amino acids, which constitute the building blocks of proteins, in rural continental precipitation. Marine and continental rainwater analyses reported by Milne and Zika [29] also indicate the presence of free and combined amino acids in the atmosphere, including those with ice-nucleating ability. Gavish [30] has demonstrated that aliphatic long-chain alcohols arranged in mono-layers at the air–water interface induce ice nucleation at temperatures as high as -0.5°C, apparently caused by a close lattice match of hexagonal ice with the alcohol crystallites, that is, two-dimensional crystalline aggregates. A similar ability can be predicted for some amino acids whose presence in the atmosphere has been demonstrated.

A widely used fertilizer, Urea or Carbamide ($NH_2CONH_2$), an organic compound synthesized from inorganic compounds, exhibits extraordinary ice nucleating properties in laboratory as well as in natural cloud [31]. As Urea is usually sprayed as fine grains, it is expected that some of this substance will go into the air as aerosol. Besides, Urea may also be considered as a good candidate for weather modification. So, study of this substance as ice nucleating agent has some relevance in practical situation as well. In fact, Knollenberg [31] first studied the ice nucleating behaviour of Urea and he reported that Urea is quite effective for ice nucleation at –10 $^0$C. Besides, one of the most surprising finding of the same experiment was that, Urea can nucleate ice up to +6 °C.

In the present work a systematic study of ice nucleation behavior with variation of seeding temperature has been undertaken in a laboratory scale cold chamber and the seeding material is taken as an aqueous solution of Urea. The temperature is being varied from –19 °C and 0 °C. Then the seeding temperature is taken to the positive side of 0 °C in order to examine if Urea solution has the capability of nucleating crystals. Incidentally, it can nucleate hydrate crystal up to +11 °C. To identify that, the crystals observed above 0 °C are really hydrate formations of Urea, a number of supporting experiments were performed. The experiments clearly identify that the concerned crystals are Urea caged by water and these are being termed as hydrate crystal. It should be noted that the aqueous solution of Ammonium Sulfate has the ability to form hydrate crystal on the positive side of 0 °C [32] and as stated earlier, Benzoin dust also has a similar behaviour [18]. Report of another hydrate crystal is also available in the available in the literature [33].

In addition, the variation of crystallization with the variation of Total Water Content (TWC) in the cold chamber has been studied keeping the seeding temperature within a narrow range of –12.0 °C to –11.0 °C.

## 2. Experimental Procedure

The laboratory observation was conducted in a single door metallic chamber where the temperature can be lowered down to –33.5 °C. The length, breadth and height of the cold chamber are 0.72 m, 0.72 m and 1.325 m respectively. Steam from a bath is allowed to pass inside the chamber through a hole near the floor of the chamber. When the steam supersaturates sufficiently the air inside the chamber at a cold temperature, super cooled droplets will appear. In the absence of any nucleus, droplets cannot freeze unless the temperature goes below - 40°C [34]. After sufficient flow of steam at cold temperatures, one can see the formation of sufficiently dense cloud inside the chamber from a glass window on the body of the chamber. The diameter of the droplets can be measured in the same way the sizes of the crystals are measured. The diameter of the droplets lies dominantly within the range of 1μm to 15 μm. The amount of water vapour in the chamber is maintained by adjusting the flow of steam. As the steam is passed, finally a steady temperature is reached. This steady temperature can be maintained for nearly five minutes. It has also been observed that this steady condition can be maintained for another two minutes at least, when the compressor of the cold chamber and the flow of steam are both cut off simultaneously. In fact, the entire nucleation experiment is performed under the later steady condition.

At the steady temperature the aqueous solution of Urea is seeded. This steady temperature is called the seeding temperature. A particular seeding temperature can be attained using two controls: fixing the temperature at which steam is allowed to pass and then, by controlling the flow of steam. 18 gm of the solute is dissolved in 50 c.c. of water and the entire amount is injected with the help of a sprayer through a seeding port near the middle of the chamber. This amount of Urea is almost at its half of the saturated amount at the room temperature. The size of the drops varies from 0.4μm to 1.0μm, with the maximum number at 0.7μm and the standard deviation is 0.21μm. In the present work, 99.5% chemically pure Urea has been used, with minor impurities being sulphate (0.001%), sulphated ash (0.01%), bluret ($C_2H_5N_3O_2$) (0.05%) with traces of chloride, iron and heavy metals apart from some trace materials. The present findings largely support the observations of Knollenberg. But, many new observations are also made, which are either unnoticed or different from or extension of those of Knollenberg. Besides, the identity of the crystals nucleated on the positive side of 0 °C is brought to light with the help of a good number of supporting experiments.

A thermistor measures the temperature near the seeding port and another thermistor is kept near the floor to measure the temperature gradient inside the chamber. The temperature difference between the two thermistors does not exceed 0.1 °C. The thermistors are calibrated at ice and steam points. The error in temperature measurement lies within 0.1 °C.

One circulator is kept fixed at the roof of the chamber to circulate the super cooled droplets evenly in the chamber and to minimize the temperature gradient within the chamber. A solution is made with 0.4 gm of formvar in 10 c.c. of Chloroform. A thin layer of this solution is coated on 4 slides at a time and these are kept near the floor to collect the signature of the crystals falling there. Immediately after the seeding, the lid over the slides is removed by outside control and the collection of crystals is carried out for one and half minutes only. So the nucleation experiment is performed in a steady condition of the environment. When the slides are brought outside after the experiment, the Chloroform quickly evaporates leaving behind a plastic film of formvar. On the other hand, ice is also converted into water. Details about the use of formvar coated slides can be found in the existing literature [35], [36]. The slides are then kept inside a desiccator with sufficient silica gel, so that water will evaporate and be quickly absorbed by silica gel. By this process a replica of the arrested crystal is retained and the nucleating agent at the core is left behind in case of heterogeneous nucleation.

One can observe the replica through a microscope. This way, the concentration of the crystals in a slide is counted and the basic habit of the crystals with change of temperature is also investigated. Here it is assumed that the crystal concentration on a slide is directly proportional to the crystal concentration in the chamber.

In order to justify that the crystals observed on the positive side of 0 $^{o}$C are really hydrated form of Urea, confirmatory tests are being performed. In the first experiment for identifying the nature of crystals formed in the positive side of 0 $^{o}$C, an over saturated solution of Urea is prepared with 75 gm of Urea being mixed with 50 ml of water. Initially the solution is heated up to 50 $^{o}$C, when the solution becomes saturated. Then the solution is kept in a beaker and allowed to cool at the room temperature. Some solute settles at the bottom, as the room temperature is less than 50 $^{o}$C. Then a slide with a small grain of Urea fixed on it is inserted within the solution. The entire system is then placed within an incubator for 48 hours keeping the temperature at a fixed value within the range 0 $^{o}$C to +15.2 $^{o}$C. Beautiful crystal growth around the bead is observed at all temperatures starting from 0 $^{o}$C, but it completely ceases at +15.2 $^{o}$C. The grown crystals were dried up and then powdered. The identification of powdered crystal was made by X-ray diffraction (XRD, BRUKER D8 ADVANCE) using $CuK_{\alpha}$ radiation. From d-values of crystal lattice, the crystal itself can be identified.

In the last confirmation experiment, the crystal grown in the incubator is subjected to $^{1}$H-NMR in $D_2O$ solvent. From chemical shift/ppm one can identify the materials present in the specimen.

## 3. Role of Urea as Ice-Nucleating Substance

In the present experiment, Urea is dissolved in water and sprayed into the chamber. It is observed that, when urea particles are dissolved in water, the temperature of the solution decreases substantially. The decrease is in the range of 10 $^{0}$C for a concentration of 18 gm of Urea in 50 c.c. of water. It is reported that the amount of absorption of heat in the solution is 252.89 x $10^3$ J/Kg of Urea. But whenever Urea solution is seeded in the chamber, the temperature of the chamber is found to fall again in the active zone of ice crystallization and then, there is a gradual rise. A detailed discussion of the temperature fall is given in section 4.4.

The temperature fall shows that during ice nucleation over Urea, there is further absorption of heat. This is quite likely, as the unbonded oxygen of Urea combines with water to form another hydrogen bonding, which again leaves open oxygen as well as another hydrogen to form further bond structure. This process continues as Urea has great affinity for continual growth. This way the entire Urea droplet forms composite packing. During these bond formations, there is overall absorption of heat.

The gradual rise in temperature after the initial fall is evidently due to release of latent heat as water undergoes a phase transition during the formation of crystal.

However Knollenberg [31] used Urea particles as seeding material and tried to explain the ice nucleation over Urea particles on the basis of Urea surface epitaxial properties.

The structure of Urea is:

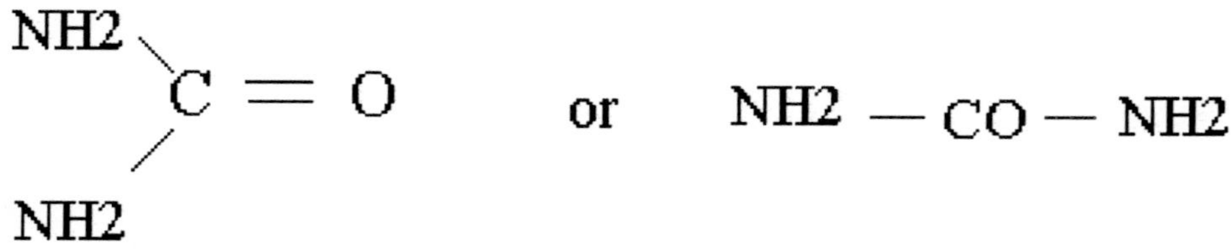


The bond length of CN is 1.47 $^{0}$A. C-N bond length in Urea has some resonance structure also:

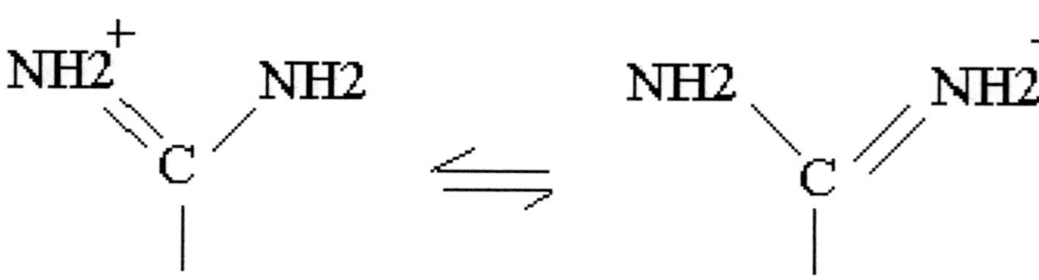


Now the negatively charged oxygen atom is capable of combining with one proton of water molecule. This leaves one O and another H open for making a new combination. This process is supposed to continue and compact packing is expected to develop with the following two structures:

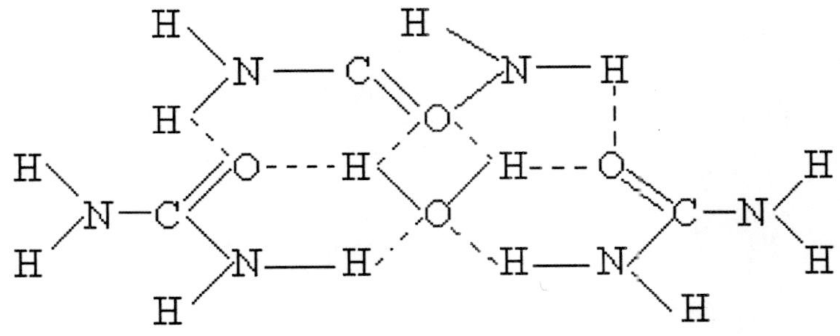

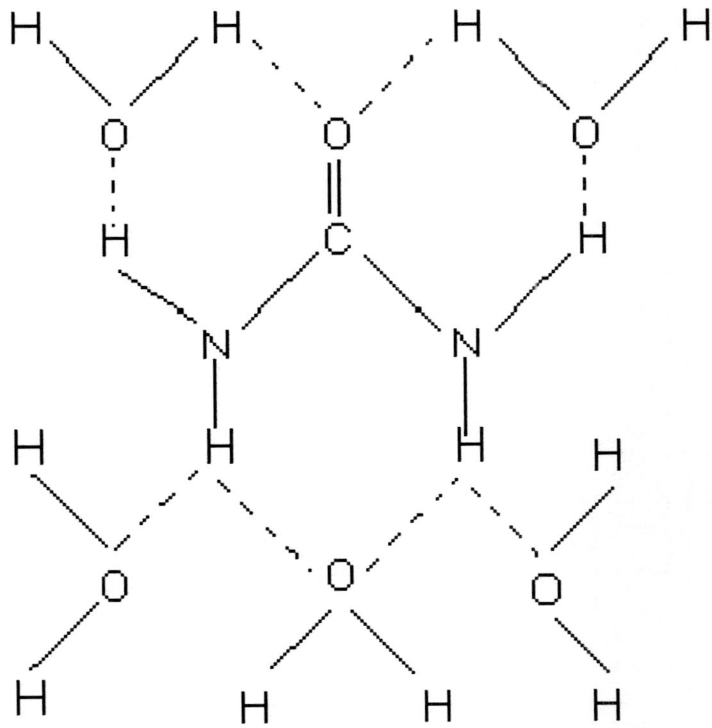

In the first structural representation, one can find that urea molecules are nucleated around a water molecule. In the second representation, water molecules are nucleated around a urea molecule. This shows that either molecule can nucleate around the other.

It should be stressed that ice crystals either in homogeneous or in heterogeneous nucleation, cannot exist above 0 $^0$C. But in heterogeneous nucleation a hydrate crystal can exist above 0 $^0$C if water can cage around the seeding nuclei such that it presents the look of a solid crystal. This is further confirmed in case of Urea by the fact that, when the slide containing the hydrate crystals is brought outside at room temperature, no water is found to come out of the crystals. This is unlike the crystals formed due to seeding on the negative side of 0 °C. Then the nucleated crystals are found to give out water as soon as the slide is brought outside at room temperature.

# 4. Results

## 4.1. Variation of Crystal Counts with Seeding Temperature

The variation of crystal count with seeding temperature is presented in Fig. 1. Here the amount of steam passed inside the chamber is kept fixed within the range of 400 gm to 500 gm. The count is made from the field of view of a microscope. In a particular slide, ten fields of view are considered. As four observations are taken at a particular temperature and there are 4 slides for one experiment, the crystal count presented in the diagram is the mean of 160 observations.

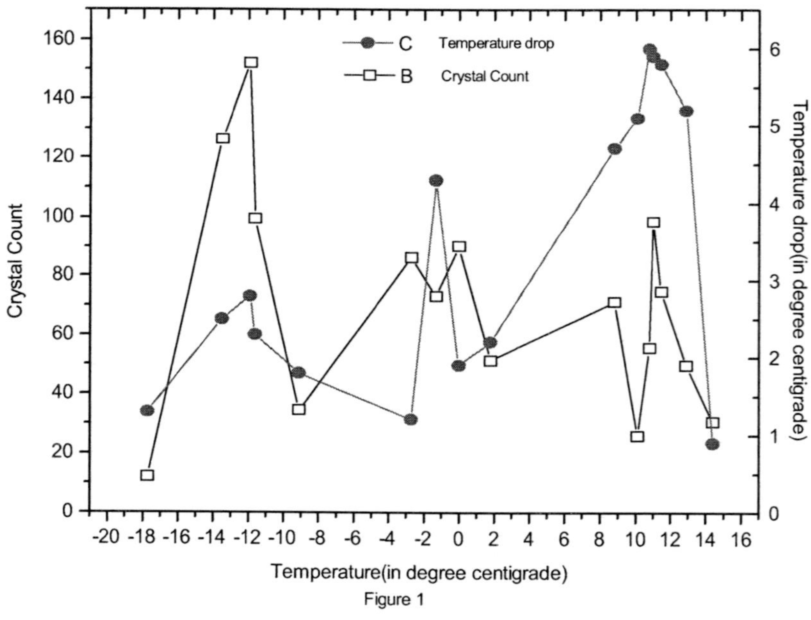


Figure 1. Seeding Temperature vs Initial Temperature Drop and Crystal count.

It is observed that the highest peak in crystal count occurs at –11.9 °C (Fig. 1). Incidentally the eutectic point for the aqueous mixture of Urea occurs around –11.5 °C [31]. It should be noted that at the eutectic temperature the Gibbs free energy heads towards a minimum which drives the system to crystallization [37]. At the lower end of temperature, the observations are taken up to –19 °C and the nucleation sharply falls there, but does not cease. The next dominant peak appears at +11.0 °C, but the nucleation definitely ceases at +15.2 °C on the higher temperature side. In between, there is another peak with some statistical fluctuations. The cut-off temperature is not presented in the Fig.1.

The most revealing part of the experiment is the existence of a significant peak in crystal count at a relatively large positive temperature. Thus the question arises – what are these crystals? In order to answer this question, a number of supporting experiments are carried out and these are being reported in 4.3.

There is still another peak close to 0 °C, but on the negative side. In the present work, no attempt is being made to understand the significance of this peak.

## 4.2. Crystal Count against TWC Variation

The observations are made around the temperature where crystal growth is maximum i.e. within the range of –11.0 $^0$C to –12.0 $^0$C. The average of more than 50 observations is presented in Fig. 2. It is observed that crystal count initially rises with TWC, but a saturation value in crystal count is observed from 0.35 Kg of TWC only. The observation is continued up to 0.65 Kg of TWC.

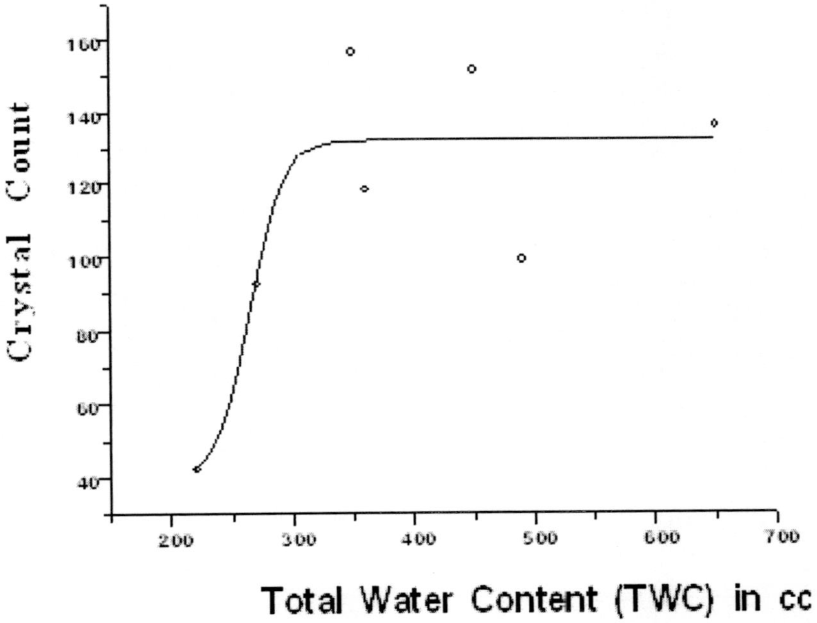


Figure 2. Crystal Count vs TWC at the temperature range of -11$^0$C to -12$^0$C.

## 4.3. Identification of Crystals Above 0 °C

When the experiment is done on the positive side of 0 °C in an incubator, beautiful growth is observed either over a slide or a rod if the temperature is kept within +15.2 °C. Selected one result of in-situ observations of growth crystals is presented in Fig. 3. At 15.2 °C, no growth is observed. The maximum growth of crystals is observed at +11 °C. In fact, the crystal growth inside the incubator at different positive temperatures is quite consistent with the observations in Fig. 1 on the positive side of 0 °C. This clearly establishes that the crystal grown in the incubator and the crystals obtained in the Cold Chamber after seeding are identical. This observation also clearly indicates the presence of hydrate crystals on the positive side of 0 °C. The image of the grown crystals is taken by a HP digital camera.

Figure 3. Growth of hydrate crystals of urea on a slide at +11 °C in an Incubator.

As the crystals grown in the incubator and in the Cold Chamber above 0 °C are being taken identical, a number of experiments are being done on the grown crystals to reveal their identity.

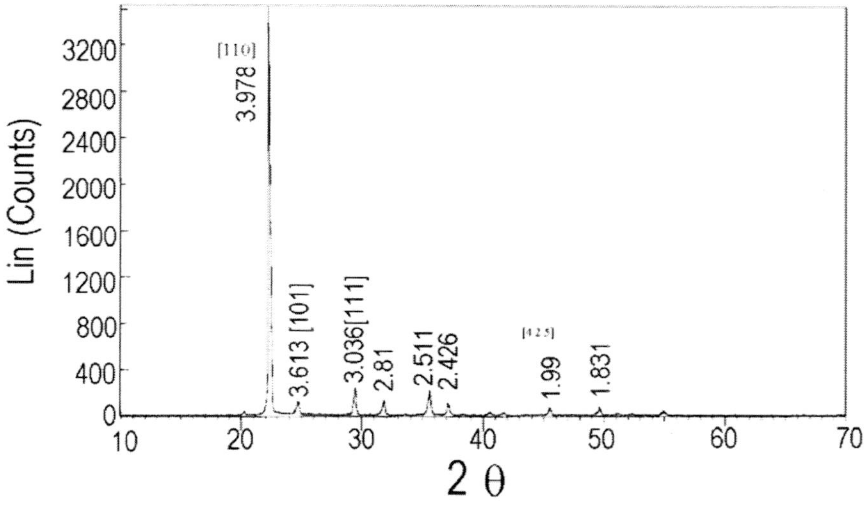


Figure 4. XRD pattern of growth crystal.

At first, X-ray diffraction (XRD) analysis of the crystal grown in the incubator is presented in Fig. 4. The pattern reveals a strong {110} orientation of Urea phase caging in water, besides the other orientations of the same crystal. The presence of hydrated Urea is also found (JCPDS-ICDD 1998-1993 [38]). However this d-value is also consistent for ice with lattice orientation {100}. Various peaks observed in XRD with 2θ-value, d-value, the possible crystal structure from JCPDS data card and the lattice orientation are presented in Table 1. From XRD analysis, one can conclude that Urea crystals are definitely present. Even in one case hydrated Urea is present. Even possibility of ice-like structure is highly probable. One should note that when Urea crystals are being caged by water to present a composite crystalline form, water behaves more like ice crystals. Possibly, that is why ice crystals are coming in the identification.

**Table 1. Observed d-values and d-values from JCPDS data card along with their {h k l}**

| Observed d-values ($d_{obs}$) | **d-values from JCPDS data card ($d_{jcpds}$)** | **{h k l}** | **2θ-value (in degree)** |
|---|---|---|---|
| 3.978 | 3.994 (37-1464)/3.91(42-1142) | {110} Urea/Ice {100} | 22.325 |
| 3.617 | 3.617 (31-1979)/3.68(16-0544) | {101} Urea/Ice {111} | 24.619 |
| 3.036 | 3.048 (31-1979)/3.03(21-0398) | {111} Urea/Ice {102} | 29.395 |
| 1.991 | 1.998 (32-0005) | {425}Al, Chloride Urea Hydrate | 45.513 |

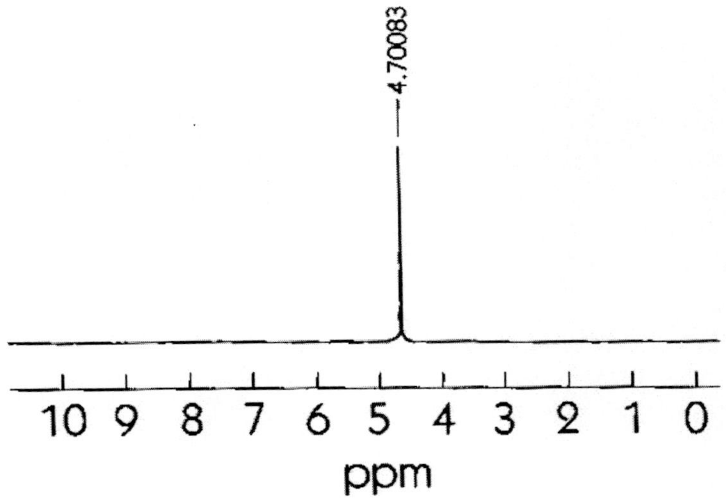


Figure 5. NMR spectrum of blank $D_2O$.

To identify the nature of grown up crystals more precisely, $^1$H-NMR study is taken up in the next. In this study, 0.5 ml of $D_2O$, which will act as solvent, is at first scanned as a standard run. A single peak of $H_2O$ is obtained at δ 4.70083 (Fig. 5). This actually gives the chemical shift/ppm for $H_2O$ present as impurity in $D_2O$. In the same $D_2O$ solvent, a small amount of grown and dried up crystal is dissolved, and the same number of scan is carried out

(Fig. 6). From shift/ppm one can identify the presence of $H_2O$ in the specimen crystal. The other peak at δ 5.68982 is due to Urea. This experiment clearly shows that only Urea and water are present in the grown up crystal.

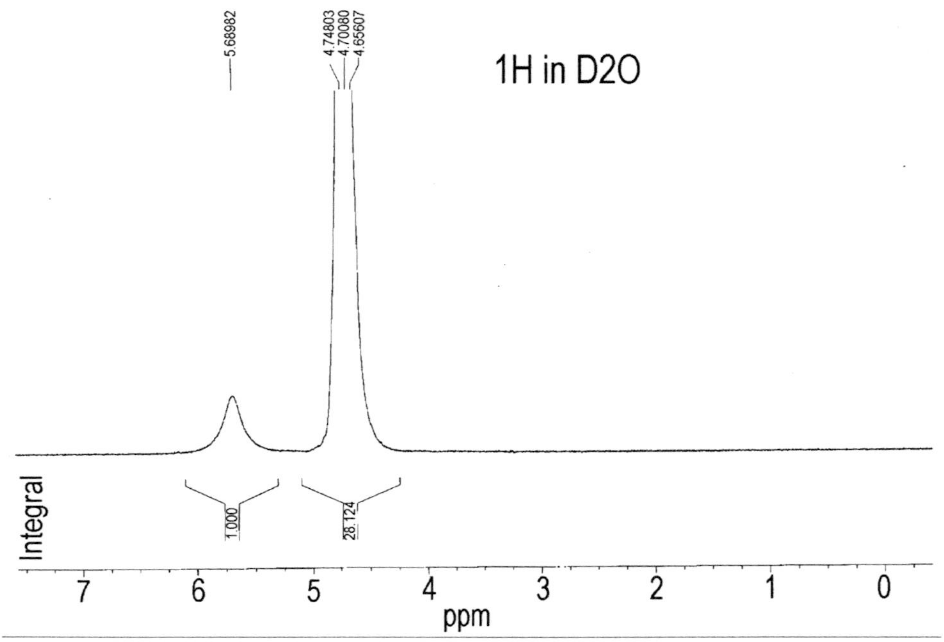


Figure 6. $^1$H-NMR spectrum of growth crystal.

To identify the nature of bonding of water in the grown up crystal, another supplementary experiment is taken up. The grown up crystal is at first dried up and then it is heated. It is observed that the crystal starts liberating water at 75 °C and by 100 °C the entire water is liberated. The entire crystal then shrinks and lumps up. In one experiment, 310 mg of dried up crystal is taken and it shrinks to 230 mg after liberation of water. The new crystal has melting point at 115 °C, whereas anhydrous Urea has melting point in the range of 132 °C -133 °C. This lowering of melting point shows that some hydrogen is still bonded up with Urea to form Urea hydrate. The melting point of Urea hydrate is lower than pure anhydrous Urea.

Now lumped up crystal is dissolved in 0.5 ml of $D_2O$ and same number of NMR scan is carried out as done in the previous experiment. From the shift/ppm study, one can conclude that no water is present in the lumped up crystal (Fig. 7). This experiment assures that the grown up crystal constitutes Urea and water, with water as only caging of Urea. As grown up crystals are considered identical with the crystals formed in the chamber, the identity of the later crystals is now better understood.

## 4.4. Variation of Initial Temperature Drop with Seeding Temperature

Variation of initial temperature drop, when Urea solution is injected, with change of seeding temperature is also presented in Fig. 1. Though the temperature drop has three peaks almost in conformity with the peaks of crystal count, there is one marked distinction. The temperature drop for the most dominant peak in crystal count is only 2.8 °C, whereas the same for second dominant peak in crystal count at 11 °C is 5.9 °C. The corresponding value for the third peak at –1.1 °C is 4.3 °C.

This clearly indicates that different processes are responsible for the different peaks in crystal count. The initial drop in temperature is certainly due to heat absorption and the explanation for the peak at –11.9 °C is already given in section 4.1.

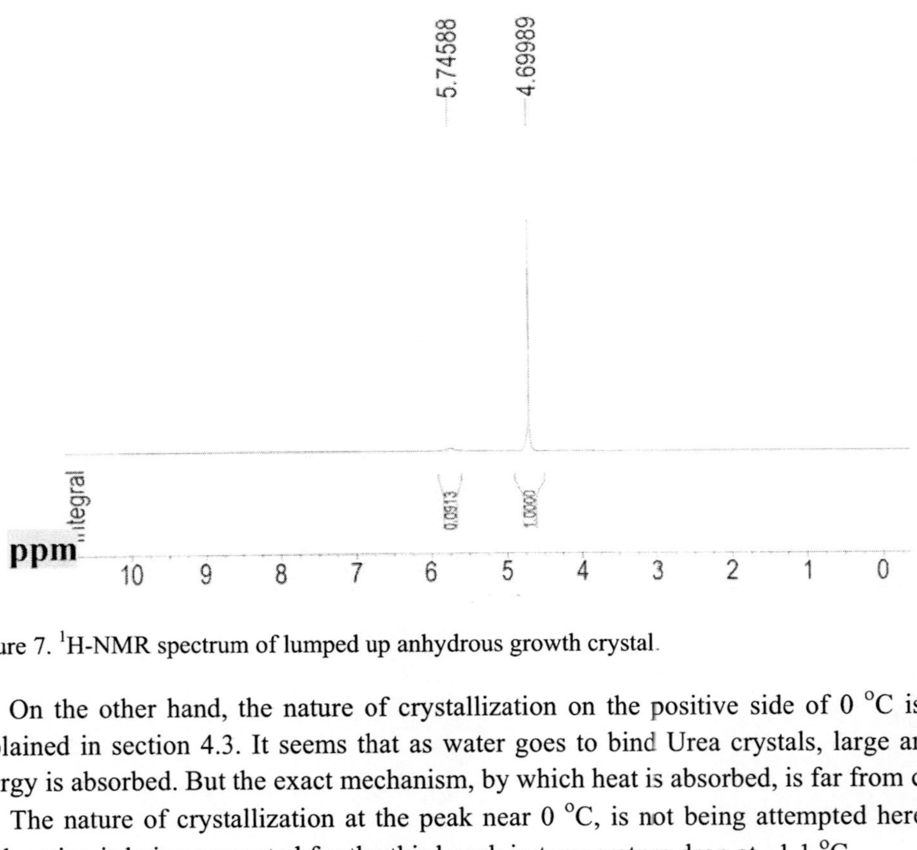


Figure 7. $^1$H-NMR spectrum of lumped up anhydrous growth crystal.

On the other hand, the nature of crystallization on the positive side of 0 °C is already explained in section 4.3. It seems that as water goes to bind Urea crystals, large amount of energy is absorbed. But the exact mechanism, by which heat is absorbed, is far from clear.

The nature of crystallization at the peak near 0 °C, is not being attempted here. So, no explanation is being presented for the third peak in temperature drop at –1.1 °C.

## 4.5. Crystal Habit at Different Temperature Zones

It is reported in the literature that the principal factor governing the basic habit of the ice crystal structure is temperature (Hanajima, [39]). Rogers and Yau [34] have also concluded that the basic habit of ice crystals is dependent on temperature and independent of seeding material. However no such conclusion can be made about the hydrated crystals. The observed variation of crystal habit with temperature is presented in Table 2.

It should be mentioned that columnar structure exists at all temperatures. But dendrite structure is never found above –16°C. Hexagonal shaped crystals are comparatively rare.

Above 0 °C, mainly rod shaped crystals are present, but the crystals are very prominent with clear edge.

**Table 2. Variation crystal morphology with variation of temperature**

| Temperature | Dominant Ice Crystal Structure |
|---|---|
| Around -20 $^0$C to -17 $^0$C | Mainly dendrite (photo 1), Rimed dendrite, Spatial dendrite, and some needle shaped. |
| -15 $^0$C to -10 $^0$C | Hexagonal (photo 2) and Prism shaped (photo 3), Rimed crystal (photo 4). |
| -10 $^0$C to - 5 $^0$C | Most of the crystals are column shaped (photo 5), Star like (photo 6). |
| -5 $^0$C to 0 $^0$C | Most of the crystals are circular shaped. |
| Above 0 °C | Very clear, distinct edged crystal with clear background and most of them are rod shaped. |

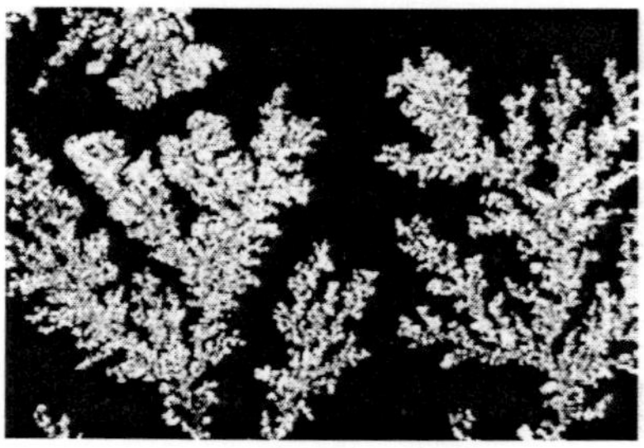

Photo 1. A dendrite structure.

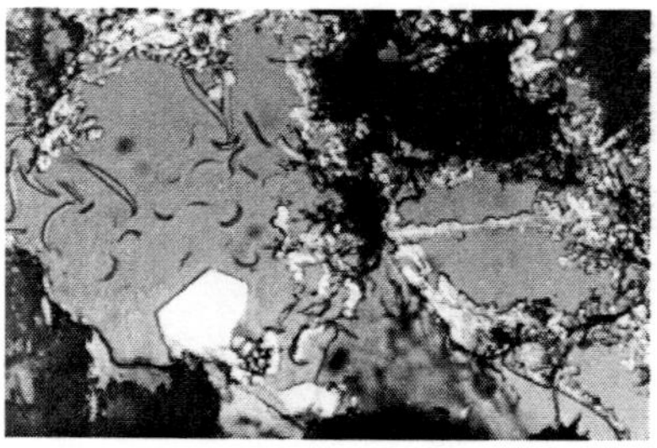

Photo 2. A one side broken hexagonal structure.

Photo 3. Prism like structure of the ice crystal

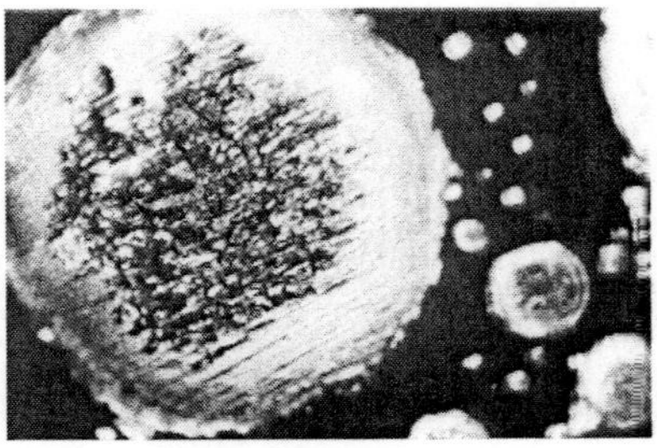

Photo 4. Bunch of rimed ice crystals

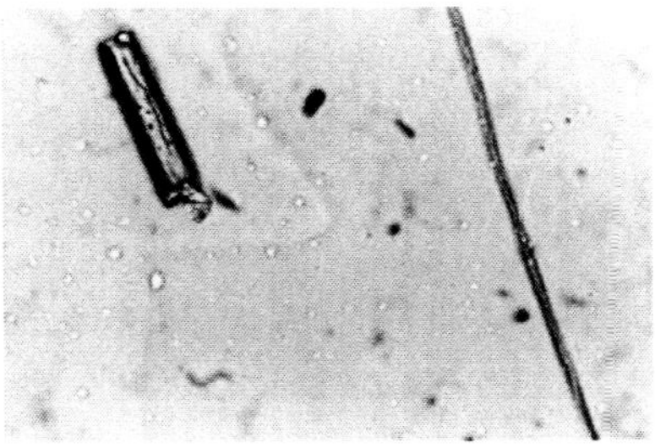

Photo 5. Rod or Column like structure.

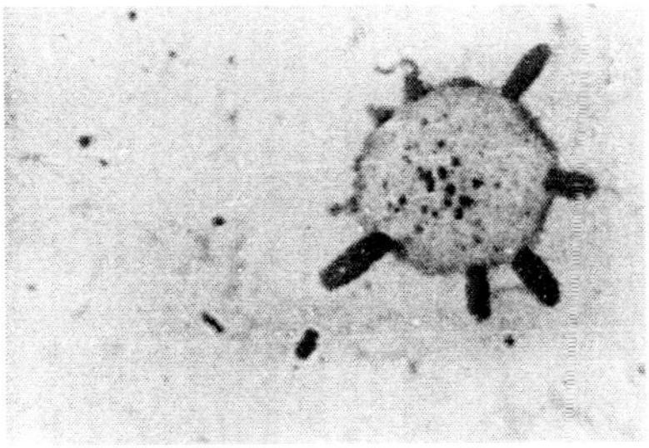

Photo 6. Star like structure of ice crystals.

## 5. CONCLUSIONS

Knollenberg [32] used urea particles mainly for ice nucleation. In the present study aqueous solution of Urea is used for the similar purpose. For the negative region of 0 °C, the present observations are not very different from his. But the present work has greatly extended the observations above 0 °C. Aqueous solution of Urea is found to nucleate ice crystal over a wide range of negative side of 0 °C. The peak in nucleation occurs at –11.9 °C, which is, close to the eutectic temperature of Urea and water mixture.

Firstly, the crystals on the positive side of 0 °C are identified as hydrate crystals of Urea with water caging around Urea. In this region, peak in crystallization occurs at +11 °C. A clear cut-off temperature for crystal growth is observed at 15.2 °C. It is concluded from different studies like crystal growth in an incubator, X-ray Diffraction pattern (XRD) and $^{1}$H-NMR analysis that hydrate crystals of Urea are formed on the positive side of 0 °C.

It can also be concluded that water molecules mainly concentrate around Urea nuclei on the positive side of 0 °C. However, the concentration of water molecules around the Urea nuclei is quite small and it helps in binding Urea molecules.

Another important observation is the existence of immediate temperature drop with seeding and its variation with seeding temperature. The three peaks in temperature drop, exactly coincide with the three peaks in crystal count. However, from change in the magnitude of temperature drop in three peak conditions, one can conclude that three different processes are active in three regions.

## 6. ACKNOWLEDGEMENT

Acknowledgement is due to India Meteorological Department; Govt. of India for the sanction of a research project and the present work forms a part of it. Thanks are also due to Dr. S. Bhar, Chemistry Department, Jadavpur University for some helpful discussions. Great appreciations are also due to M/S Ruoss Aircondition Co., Kolkata for the erection of the equipment for the experiment.

## 7. REFERENCES

[1] H. R. Pruppacher, J. D. Klett, *Microphysics of Clouds and Precipitation*, Kluwer Academic Publishers (Dordrecht) 1998.
[2] B. Vonnegut, *J. Appl. Phys*., 1947, vol, 18, 593.
[3] M. Kumai, M. Francis, *J. Atmos. Sci.,* 1962, vol, 19, 434.
[4] K. Isono, *J. Meteorl.,* 1955, vol, 12, 456.
[5] J. Rucklidge, *J. Atmos. Sci*., 1965, vol, 22 , 301.
[6] K. Isono, M. Komabayasi, A. Ono, *J. Meteorl. Soc., Japan*, 1959, vol, 37, 211.
[7] P. V. Hobbs, G. C. Bluhm, T. Ohtake, *Tellus*, 1971, vol, 23, 28.
[8] N. Fukuta, *J. Meteorl.,* 1958, vol, 15, 17.
[9] R. Serpolay, *Bull. Obs. Puy de Dome*, 1958, No.3, 81.
[10] B. J. mason, H. P. Van den Heuvel, *Proc. Phys. Soc., London*, 1959, 74A, 744.
[11] U. Katz, *Z. Angew. Math. Phys*., 1960, vol, 11, 237.
[12] R. F. Pueschel, *J. Wea. Mod*., 1978, vol, 10, 9.
[13] J. Hallett and C. Knight, *Atmos. Res*., 1994, vol, 32, 1-11.
[14] R. R. Braham (Jr.), *Amer. Meteor. Soc.,* 1986, no, 43, 15.
[15] G. Langer, J. Rosinski, and N.S. Berusen, *J.Atmos.Sci*., 1963, vol, 20, 557-562.
[16] R. R. Braham, *J. Atmos. Sci*., 1963, vol, 20, 563-567.
[17] S. Paul, A. Hazra, D. Roy, U. K. De, S. Bhar and K. Goswami, *J. Weather Modification*, 2004, vol, 36, 41-46.
[18] S. Paul, A. Hazra, U. K. De, S. Bhar and K. Goswami, *14th ICPP conference proceedings*, 2004, vol, 2, 1257-1260.
[19] M. Komabayashi and Y. Ikebe, *J. Meteor. Soc. Japan,* 1961, vol, 39, 82.
[20] R. B. Head, *Nature*, 1961, vol, 191, 1058.
[21] V. A. Garten, and R. B. Head, *Nature*, 1965, vol, 205, 160.

[22] N. Fukuta, *J. Atmos., Sci*, 1966, vol, 23, 191.
[23] W. Szyrmer, I. Zawadski, *Bull. Am. Meteorl. Soc.*, 1997, vol,78, 209.
[24] B. A. Power, and R. F. Power, *Nature*, 1962, vol, 194, 1170.
[25] S. Matthias-Maser, R. Jaenicke, *Atmos. Res.*, 1995, vol, 39, 279.
[26] N. Barthakur, and J. Maybank, *Nature*, 1963, vol, 200, 866.
[27] N. Barthakur, and J. Maybank, *J. Rech. Atmos.*, 1966, vol, 23, 475.
[28] D. G. Georzelskopolus and D. C. Sands, *Can. J. Microbiol.*, 1992, vol, 38, 111.
[29] P. J. Milne and R. G. Zika, *J. Atmos., Chem.*, 1963, vol, 16, 361.
[30] M. Gavish, R. Popovitz-Biro, M. Lahav, and L. Leiserowitz, *Science*, 1990, vol, 250, 973-974.
[31] R.G. Knollenberg, *J.Atmos.Sci.,* 1966, vol, 23, 197-201.
[32] Hazra, S. Paul, U. K. De, S. Bhar and K. Goswami, *Progress in Crystal Growth and Characterization of Materials*, 2004, vol, 47, 45-61.
[33] J. A. Ripmeester, J. S. Tse, C. I. Ratclifffe, B. M. Powell, *Nature*, 1987, vol, 325, 135.
[34] R.R. Rogers, and M.K. Yau, *A short course in Cloud Physics*. (Third edition), Pergamon Press (Oxford), 1989, pp 156.
[35] P. R. Saunders and N. A. Wahab, *J. Appl. Meterol.*, 1973, vol, 12, 1035.
[36] J. Hallet, *Operating Manual*, Air Force Geophysics Laboratory, United States Air Force, Hanscom AFB, Mass., 1970.
[37] L. V. Azaroff, *Introduction to Solids*, McGraw-Hill (New York), 1995, pp 188-189.
[38] JCPDS-ICDD 1988-1993, *JCPDS- International Center for Diffraction Data*, 12 campus Blvd. NS, PA 19073-3273 U.S.A.
[39] M. Hanajima, *Low Temp. Sci.*, 1954, vol, A1, 53-65.

# INDEX

## A



## B



## C

## D



## E



## F



## G



## H

## I



## K



## L



## M



## N



## O

## P



## Q



## R



## S



## T